READY-TO-USE

EARTH & ASTRONOMICAL SCIENCE ACTIVITIES

FOR GRADES 5-12

Mark J. Handwerker, Ph.D.

THE CENTER FOR APPLIED
RESEARCH IN EDUCATION
West Nyack, New York 10994

Library of Congress Cataloging-in-Publication Data

Handwerker, Mark J.
    Ready-to-use earth & astronomical science activities for grades 5–12 / Mark J.
Handwerker, Ph.D.
        p.   cm.—(Secondary science curriculum activities library)
    ISBN 0-87628-445-4
    1. Geophysics—Study and teaching (Secondary)—Curricula.   2. Astronomy—
Study and teaching (Secondary)—Curricula.   3. Curriculum planning.   4. Astronomy
teachers—Handbooks, manuals, etc.   5. Geology teachers—Handbooks, manuals, etc.
I.   Title.   II.   Title: Ready-to-use earth and astronomical science activities for grades 5–12.
III. Title: Earth & astronomical science activities for grades 5–12.   IV. Title: Earth and
astronomical science activities for grades 5–12.   V. Series.
Qc807.5.H36   1999
550'.71'2—dc21   98-48063
                                                                                                    CIP

© 1999 *by* The Center for Applied Research in Education, West Nyack, NY

Printed in the United States of America

10   9   8   7   6   5   4   3   2

ISBN 0-87628-445-4

ATTENTION: CORPORATIONS AND SCHOOLS
The Center for Applied Research in Education books are available at quantity discounts with
bulk purchase for educational, business, or sales promotional use. For information, please
write to: Prentice Hall Direct Special Sales, 240 Frisch Court, Paramus, NJ 07652. Please
supply: title of book, ISBN number, quantity, how the book will be used, date needed.

**THE CENTER FOR APPLIED RESEARCH
IN EDUCATION**
West Nyack, NY 10994

On the World Wide Web at http://www.phdirect.com

About This Resource

Ready-to-Use Earth & Astronomical Science Activities for Grades 5–12 is designed to help you teach basic science concepts to your students while building their appreciation and understanding of the work of generations of curious scientists. Although The Scientific Method remains the most successful strategy for acquiring and advancing the store of human knowledge, science is—for all its accomplishments—still merely a human endeavor. While the benefits of science are apparent in our everyday lives, its resulting technology could endanger the survival of the species if it is carelessly applied. It is therefore essential that our students be made aware of the nature of scientific inquiry with all its strengths and limitations.

A primary goal of science instructors should be to make their students "science literate." After completing a course of study in any one of the many scientific disciplines, students should be able to:

1. appreciate the role played by observation and experimentation in establishing scientific theories and laws,

2. understand cause-and-effect relationships,

3. base their opinions on fact and observable evidence—not superstitions or prejudice, and

4. be willing to change their opinions based on newly acquired evidence.

Scientific theories come and go as new observations are made. During the course of instruction, teachers should emphasize the "process" of science as well as the relevance of pertinent facts.

This volume of science activities was designed to accomplish all of the above, keeping in mind the everyday challenges faced by classroom instructors.

On Your Mark!

Begin by stimulating students' gray matter with basic scientific concepts through brainstorming and open discussion.

Get Set!

Kindle interest by making concepts real through demonstration and/or descriptive analogy.

Go!

Cement concepts into concrete form with exciting hands-on experience.

Each of the 15 teaching units in this volume of *Ready-to-Use Earth & Astronomical Science Activities for Grades 5–12* contains *four* 40–50 minute lessons and follows the same instructional sequence so that your students will always know what is expected of them. Each unit comes complete with the following:

- a **Teacher's Classwork Agenda for the Week** and **Content Notes for Lecture and Discussion,**
- a student **Fact Sheet** with **Homework Directions** on the back,
- four 40–50 minute **Lesson Plans,** each followed by its own **Journal Sheet** to facilitate student notetaking, and
- an end-of-the-unit **Review Quiz.**

Each unit has been tested for success in the classroom and is ready for use with minimal preparation on your part. Simply make as many copies of the Fact Sheet with Homework Directions, Journal Sheets, and Review Quizzes as you need for your class. Also, complete answer keys for the homework assignments and unit quiz are provided at the end of the Teacher's Classwork Agenda for the Unit.

Mark J. Handwerker

ABOUT THE AUTHOR

Mark J. Handwerker (B.S., C.C.N.Y., Ph.D. in Biology, U.C.I.) has taught secondary school science for 15 years in the Los Angeles and Temecula Valley Unified School Districts. As a mentor and instructional support teacher, he has trained scores of new teachers in the "art" of teaching science. He is the author/editor of articles in a number of scientific fields and the coauthor of an earth science textbook (Harcourt Brace Jovanovich, *Earth Science*) currently in use.

Dr. Handwerker teaches his students that the best way to learn basic scientific principles is to become familiar with the men and women who first conceived them. His classroom demonstrations are modeled on those used by the most innovative scientists of the past. He believes that a familiarity with the history of science, and an understanding of the ideas and methods used by the world's most curious people, are the keys to comprehending revolutions in modern technology and human thought.

Suggestions for Using These Science Teaching Units

The following are practical suggestions for using the 15 teaching units in this resource to maximize your students' performance.

Fact Sheet

At the start of each unit, give every student a copy of the **Fact Sheet** for that unit with the **Homework Directions** printed on the back. The Fact Sheet introduces content vocabulary and concepts relevant to the unit. You can check students' homework on a daily basis or require them to manage their own "homework time" by turning in all assignments at the end of the unit. Most of the homework assignments can be completed on a single sheet of standard-sized (8½" × 11") looseleaf paper. Urge students to take pride in their accomplishments and do their most legible work at all times.

Journal Sheet

At the start of each lesson, give every student a copy of the appropriate **Journal Sheet** which they will use to record lecture notes, discussion highlights, and laboratory activity data. Make transparencies of Journal Sheets for use on an overhead projector. In this way, you can model neat, legible, notetaking skills.

Current Events

Since science does not take place in a vacuum (and also because it is required by most State Departments of Education), make **Current Events** a regular part of your program. Refer to the brief discussion on "Using Current Events to Integrate Science Instruction Across Content Areas" in the Appendix.

Review Quiz

Remind students to study their Fact and Journal Sheets to prepare for the end-of-the-unit **Review Quiz**. The Review Quiz is a 15-minute review and application of unit vocabulary and scientific principles.

Grading

After completing and collectively grading the end-of-the-unit Review Quiz in class, have students total their own points and give themselves a grade for that unit. For simplicity's sake, point values can be awarded as follows: a neatly completed set of Journal Sheets earns 40 points; a neatly completed Homework Assignment earns 20 points; a neatly completed Current Event earns 10 points; and, a perfect score on the Review Quiz earns 30 points. Students should record their scores and letter grades on their individual copies of the **Grade Roster** provided in the Appendix. Letter grades for each unit can be earned according to the following point totals: A ≥ 90, B ≥ 80, C ≥ 70, D ≥ 60, F < 60. On the reverse side of the Grade Roster, students will find instructions for calculating their "grade point average" or "GPA." If they keep track of their progress, they will never have to ask "How am I doing in this class?" They will know!

Unit Packets

At the end of every unit, have students staple their work into a neat "unit packet" that includes their Review Quiz, Homework, Journal Sheet, Current Event, and Fact Sheet. Collect and examine each student's packet, making comments as necessary. Check to see that students have awarded themselves the points and grades they have earned. You can enter individual grades into your record book or grading software before returning all packets to students the following week.

You will find that holding students accountable for compiling their own work at the end of each unit instills a sense of responsibility and accomplishment. Instruct students to show their packets and Grade Roster to their parents on a regular basis.

Fine Tuning

This volume of *Ready-to-Use Earth & Astronomical Science Activities for Grades 5–12* was created so that teachers would not have to "reinvent the wheel" every week to come up with lessons that work. Instructors are advised and encouraged to fine tune activities to their own personal teaching style in order to satisfy the needs of individual students. You are encouraged to supplement lessons with your district's adopted textbook and any relevant audiovisual materials and computer software. Use any and all facilities at your disposal to satisfy students' varied learning modalities (visual, auditory, kinesthetic, and so forth).

CONTENTS

EA1   MAPPING THE EARTH / 1

Teacher's Classwork Agenda and Content Notes

Classwork Agenda for the Week . . . Content Notes for Lecture
and Discussion . . . Answers to the End-of-the-Week Review Quiz

Fact Sheet with Homework Directions

Lesson #1
Students will explain how maps are used to determine locations on a globe.
Journal Sheet #1

Lesson #2
Students will use a compass and sundial to tell time and direction.
Journal Sheet #2

Lesson #3
Students will use a local map to determine the distance between points of interest.
Journal Sheet #3

Lesson #4
Students will draw a contour map and construct a corresponding relief model.
Journal Sheet #4

EA1 Review Quiz

EA2   THE STRUCTURE OF THE EARTH / 15

Teacher's Classwork Agenda and Content Notes

Classwork Agenda for the Week . . . Content Notes for Lecture
and Discussion . . . Answers to the End-of-the-Week Review Quiz

Fact Sheet with Homework Directions

Lesson #1
Students will show how earthquake waves help us to determine the earth's internal structure.
Journal Sheet #1

Lesson #2
Students will discuss evidence that earth's continents have moved across the planet's surface.
Journal Sheet #2

Lesson #3
Students will show how moving crustal plates build mountains and ocean trenches.
Journal Sheet #3

Lesson #4
Students will show how heat from the earth's interior produces geysers and volcanoes.
Journal Sheet #4

EA2 Review Quiz

EA3   CHANGING LAND / 29

Teacher's Classwork Agenda and Content Notes

Classwork Agenda for the Week . . . Content Notes for Lecture and Discussion . . . Answers to the End-of-the-Week Review Quiz

Fact Sheet with Homework Directions

Lesson #1
Students will identify the major forces that change the surface of earth's crust.
Journal Sheet #1

Lesson #2
Students will examine the effects of wind and water erosion.
Journal Sheet #2

Lesson #3
Students will examine the effects of mechanical and chemical weathering.
Journal Sheet #3

Lesson #4
Students will discuss how glaciers have changed the face of continents.
Journal Sheet #4

EA3 Review Quiz

EA4   ROCKS AND MINERALS / 43

Teacher's Classwork Agenda and Content Notes

Classwork Agenda for the Week . . . Content Notes for Lecture and Discussion . . . Answers to the End-of-the-Week Review Quiz

Fact Sheet with Homework Directions

Lesson #1
Students will explain the rock cycle.
Journal Sheet #1

Lesson #2
Students will examine the basic characteristics of igneous rocks.
Journal Sheet #2

Lesson #3
Students will examine the basic characteristics of sedimentary rocks.
Journal Sheet #3

Lesson #4
Students will examine the basic characteristics of metamorphic rocks.
Journal Sheet #4

EA4 Review Quiz

EA5   EARTH ORIGINS AND GEOLOGIC TIME / 57

Teacher's Classwork Agenda and Content Notes

Classwork Agenda for the Week . . . Content Notes for Lecture
and Discussion . . . Answers to the End-of-the-Week Review Quiz

Fact Sheet with Homework Directions

Lesson #1
Students will explain how gravity caused the formation of the earth and solar system.
Journal Sheet #1

Lesson #2
Students will examine how to determine the age of rocks and rock strata.
Journal Sheet #2

Lesson #3
Students will create a geologic time chart of the Cryptozoic Eon.
Journal Sheet #3

Lesson #4
Students will create a geologic time chart of the Phanerozoic Eon.
Journal Sheet #4

EA5 Review Quiz

EA6   THE OCEANS / 71

Teacher's Classwork Agenda and Content Notes

Classwork Agenda for the Week . . . Content Notes for Lecture
and Discussion . . . Answers to the End-of-the-Week Review Quiz

Fact Sheet with Homework Directions

Lesson #1
Students will show how sonar is used to determine the depth of the oceans.
Journal Sheet #1

Lesson #2
Students will map the patterns of the world's major ocean currents.
Journal Sheet #2

Lesson #3
Students will explain the causes of waves and ocean tides.
Journal Sheet #3

Lesson #4
Students will list and describe the components of seawater.
Journal Sheet #4

EA6 Review Quiz

EA15 STARS AND GALAXIES
OF THE COSMOS / 197

Teacher's Classwork Agenda and Content Notes

**Classwork Agenda for the Week . . . Content Notes for Lecture
and Discussion . . . Answers to the End-of-the-Week Review Quiz**

Fact Sheet with Homework Directions

Lesson #1
Students will describe the methods scientists use to measure the distance to faraway stars.
Journal Sheet #1

Lesson #2
Students will describe the forces that give birth to stars and galaxies.
Journal Sheet #2

Lesson #3
Students will describe the forces that cause stars to age and die.
Journal Sheet #3

Lesson #4
Students will examine evidence that the universe is expanding.
Journal Sheet #4

EA15 Review Quiz

APPENDIX / 211

John Couch Adams
Buzz Aldrin
Luis Alvarez
Archimedes
Aristotle
Neil Armstrong
Robert D. Ballard
Francis Beaufort
Vilhelm F.K. Bjerknes
Johann Elert Bode
Tycho Brahe
Werner von Braun
Alexandre Brongniart
Robert Bunsen
Anders Celsius
Nicolaus Copernicus
Gaspard Gustave de Coriolis
Nicholas de Cusa
Georges Cuvier
Charles Darwin
William Morris Davis
René Descartes
Cornelis Drebbel
Albert Einstein
Eratosthenes
Euclid
Gabriel Daniel Fahrenheit
Jean Bernard Léon Foucault
Joseph von Fraunhofer
Yuri Gagarin
Galileo Galilei
Johann Gottfried Galle
George Gamow

Hans Geiger
Robert Hutchings Goddard
Beno Gutenburg
George Hadley
Edmond Halley
John Harrison
René-Just Haüy
Joseph Henry
Sir William Herschel
Ejnar Hertzsprung
Hipparchus
Arthur Holmes
Robert Hooke
Edwin Powell Hubble
James Hutton
Immanuel Kant
William Thomson Kelvin
Johannes Kepler
Gustav Robert Kirchoff
Wladmir Peter Köppen
Paul Langevin
Henrietta Swan Leavitt
Inge Lehmann
Georges Edouard LeMaître
Hans Lippershey
Percival Lowell
Charles Lyell
Gerhard Kremer Mercator
Milutin Milankovich
John Milne
Friedrich Moh
Joseph Michel & Jacques
  Etienne Montgolfier

Samuel F.B. Morse
John Murray
Sir Isaac Newton
Richard Dixon Oldham
Abraham Ortelius
Auguste Antoine & Jacques
  Piccard
William Henry Pickering
Plato
Joseph Priestley
Claudius Ptolemaeus
William Redfield
Sally Kirsten Ride
Henry Norris Russell
Karl Wilhelm Scheele
Adam Sedgwick
Alan Bartlett Shepard
William Smith
Nikolaus Steno
Valentina V. Tereshkova
Johann Daniel Titius
Clyde William Tombaugh
Evangelista Torricelli
Konstantin E. Tsiolkovsky
James Alfred van Allen
Jules Verne
Alfred Lothar Wegener
Abraham Gottlob Werner
Fred Lawrence Whipple
William Wollaston
Orville & Wilbur Wright

MAPPING THE EARTH

TEACHER'S CLASSWORK AGENDA AND CONTENT NOTES

Classwork Agenda for the Week

1. Students will explain how maps are used to determine locations on a globe.
2. Students will use a compass and sundial to tell time and direction.
3. Students will use a local map to determine the distance between points of interest.
4. Students will draw a contour map and construct a corresponding relief model.

Content Notes for Lecture and Discussion

The science of **geology**, or study of the earth, depends on the use of maps. Accurate **topographic maps** that show the relief of a region, and **geologic maps** which show the kinds of rocks and land features present in an area, are the creation of skilled surveyors and navigators. Although the oldest maps date back to the Babylonians in 2300 B.C., the art and science of mapmaking began to flourish in 15th century Europe. Exploration of the New World demanded better navigational tools and skilled draftsmen such as Netherland cartographers **Gerhard Kremer Mercator** (b. 1512; d. 1594) and **Abraham Ortelius** (b. 1527; d. 1598) who published the first modern atlas. The invention of the **compass** for determining direction, the **sextant** for determining latitude, and the **marine chronometer** for determining longitude on long sea voyages greatly improved the art of navigation in the 16th and 17th centuries. Both the ancient Greeks (as early as 500 B.C.) and the Chinese (around 1088 A.D.) made use of simple magnetic compasses. An early form of the sextant, called an octant, was produced in 1731 by the English inventor **John Hadley** (b. 1682; d. 1744). And the first marine chronometer was invented in England in 1735 by **John Harrison** (b. 1693; d. 1776). All of these instruments facilitated the exploration and circumnavigation of the globe. Today, maps are largely transformations of aerial and satellite photography.

Mapmaking involves five basic ingredients: *projection, scale, orientation, legend,* and *topography.* A **projection** is a two-dimensional representation of a three-dimensional surface that allows land shapes to be displayed in their proper proportion. The **Mercator, polar** and **polyconic** projections shown in Figure A are the most frequently used. Although valuable for its accurate depiction of land shapes at the equator, a Mercator projection distorts land shapes at higher latitudes. A polar projection corrects for distortion at polar latitudes. The polyconic projection is the most useful for depicting small regions of the earth's surface (e.g., a continent) with little or no distortion. Maps are **scaled** to show the proper relationship between the sizes of land features, since a map is much smaller than the actual land region. Written or verbal, graphic, and fractional scales are discussed in Lesson #3 of this unit. All maps are **oriented** to show the four basic directions: north, south, east,

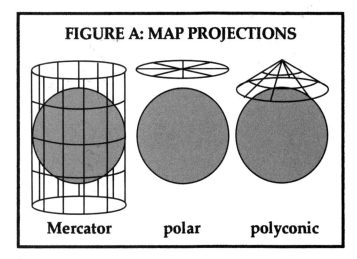

FIGURE A: MAP PROJECTIONS

Mercator   polar   polyconic

and west. And, **legends** are employed to summarize the descriptive symbols used in the interpretation of natural (e.g., **topographic**) and manmade features present in a region.

In Lesson #1, students will learn to distinguish **latitude** from **longitude** and be able to explain how maps are used to determine locations on a globe.

In Lesson #2, students will use a compass and sundial to tell direction and time.

In Lesson #3, students will learn how to interpret map scales and use a local map to determine the distance between points of interest.

In Lesson #4, students will draw a contour map and construct a corresponding relief model to see how three-dimensional surfaces can be depicted in two dimensions.

ANSWERS TO THE HOMEWORK PROBLEMS

Student maps will vary but should indicate that the student has a grasp of proper orientation and scale.

ANSWERS TO THE END-OF-THE-WEEK REVIEW QUIZ

1. model	6. compass	11. parallels	16. shaded near middle
2. true	7. meridians	12. equator	17. east face
3. true	8. intersect at	13. true	18. west face
4. true	9. Prime meridian	14. true	19. east
5. true	10. true	15. geologist	20. left drawing shaded

EA1 FACT SHEET

MAPPING THE EARTH

CLASSWORK AGENDA FOR THE WEEK

(1) Explain how maps are used to determine locations on a globe.
(2) Use a compass and sundial to tell direction and time.
(3) Use a local map to determine the distance between points of interest.
(4) Draw a contour map and construct a corresponding relief model.

A **map** is a model of the surface of the earth. The first world maps were made by the Babylonians in Mesopotamia (known today as the Middle East) around 2300 B.C. Babylonian maps were made of clay. And Babylon was placed at the center of the map since little was known of the rest of the world at the time. Around 150 A.D., the Egyptian astronomer **Claudius Ptolemaeus** (a.k.a. **Ptolemy**; b. 100 A.D.; d. 170 A.D.), developed the theory that the earth was the center of the universe. He drew valuable maps of Europe, North Africa, and the Middle East which served as guides to explorers for the next 1,500 years. Educated people of Ptolemy's time knew that the earth was a sphere. However, they could not figure out how to draw an accurate two-dimensional map of a globe which is a three-dimensional object. In 1569, the Flemish mapmaker **Gerhard Kremer** (a.k.a. **Mercator**; b. 1512; d. 1594) published the first "flat" map showing accurately plotted points taken from a curved surface.

Over the centuries, **navigators** have found their way around the planet by land or sea using maps. Of course, a map is useless unless it is placed in the correct position compared to a known direction. For those travelling at night, the **North Star** is used as a common point of reference when exploring the Northern Hemisphere. The North Star is always in a "fixed" position in the sky relative to other stars. In the Southern Hemisphere travelers use the **Southern Cross** as their point of reference. The Southern Cross, a familiar grouping of stars to people living in the Southern Hemisphere, appears on the national flag of Australia. But how did ancient explorers navigate during the day? Around 500 B.C., the ancient Greeks noticed that a piece of **lodestone** hung from a string always pointed toward the North Star. Since the magnetized metal always pointed north, they could navigate during the day without the aid of the stars. They had invented the **compass**. The Chinese astronomer and mathematician **Shen Kua** (b. 1031; d. 1095) placed a **magnetized needle** through a piece of straw and floated the straw in a bowl of water. His compass also pointed in the north-south direction.

Locations on a map are found using *meridians* and *parallels*. **Meridians** are drawn vertically on a map with the north direction positioned at the top of the map. On a globe, all meridians cross at the North and South Poles. Examine Figure I. Meridians called **longitude lines** are used to find positions east or west of the **Prime Meridian**. In 1884, astronomers chose the small town of Greenwich, England, to be the Prime Meridian or 0° longitude. On the opposite side of the globe, around 180° longitude, lies the **International Date Line**. As the earth rotates on its axis, the new day begins at midnight at the International Date Line. **Parallels** are drawn horizontally on a map, west to east across the map. Parallels, also called **latitude lines**, are "perpendicular" to meridians. A latitude line measures distances north or south of the earth's equator. The **equator** is an imaginary latitude line, 0° latitude, that runs around the middle circumference of the earth. The North and South Poles are at 90° latitude. Since the earth is about 24,000 miles (38,700 kilometers) in circumference, a single degree in latitude or longitude represents a distance of about 67 miles (108 km) at the equator.

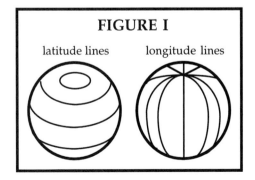

FIGURE I

latitude lines   longitude lines

EA1 Fact Sheet (cont'd)

Earth scientists called **geologists** use two basic types of maps: **topographic maps** and **geologic maps**. A topographic map shows the position, size, shape, and physical features of an area. A topographic map shows elevations above or below sea level using **contour lines**. A geologic map indicates the types of rocks and landforms in a given area. Geologic maps are used by scientists to keep a record of mineral, ore, and fossil deposits.

Homework Directions

Draw a map of your bedroom. Find out how your bedroom is positioned relative to the North Star using a compass or by noting the position of the North Star at night. Draw parallels and meridians on your map and use an appropriate "scale" (i.e., 1 inch : 1 foot) to show the position of objects in your room. Refer to your notes on Journal Sheet #1 to help you.

Assignment due: _____

| _____ | _____ | ____/____/____ |
| Student's Signature | Parent's Signature | Date |

MAPPING THE EARTH

Work Date: ____/____/____

LESSON OBJECTIVE

Students will explain how maps are used to determine locations on a globe.

Classroom Activities

On Your Mark!

Attempting to draw the curved surface of the earth on a flat piece of paper invariably results in distortion. However, three-dimensional landforms can be accurately translated into two-dimensional form by using a mathematical device called a **projection**. Draw Illustration A on the board and have students copy your notes on Journal Sheet #1. Have them imagine that a piece of photographic film has been wrapped around a globe to form a cylinder and that the surface of the globe is transparent except for the continents which are opaque. A light source at the center of the globe exposes the film. Point out how Continent A is distorted less than Continent B because A lies nearer the equator. Continent B is highly distorted because it lies nearer the pole. The map created with this methed is called a **Mercator projection** after **Gerhard Kremer Mercator** (b. 1512; d. 1594). Display a world map shown in Mercator projection and compare it to a globe of the earth. Show students how Greenland, the world's largest island, seems as large as continents such as Australia. In reality, Greenland is much smaller.

Get Set!

Have students refer to the Fact Sheet for an explanation of **parallels** and **latitude lines**, **meridians** and **longitude lines**. Have students copy your drawing of Illustration B to see how a globe can also be viewed from the "top" in a **polar projection**. Have students begin the activity described in Figure A on Journal Sheet #1 by using a felt marker to draw parallels, meridians, and islands on an inflated, round balloon. Have them choose a "prime meridian" and list the latitude and longitude of their scattered islands.

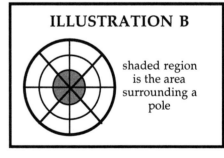

ILLUSTRATION A

globe with continents wrapped in a cylinder of photographic film

section of film

projection of Continent "B"

projection of Continent "A"

globe

ILLUSTRATION B

shaded region is the area surrounding a pole

Go!

When students are finished plotting the latitude and longitude of their islands, have them deflate their balloon and cut along the prime meridian half way around their planet. Have them attempt to stretch the balloon out as best they can into a rectangle. Point out how the parallels and meridians in particular change shape when one attempts to flatten the map into a two-dimensional sheet.

Materials

Mercator map of the world, a globe, round balloons, felt tip markers, scissors

Name: _____   Period:_____   Date: ____/____/____

MAPPING THE EARTH

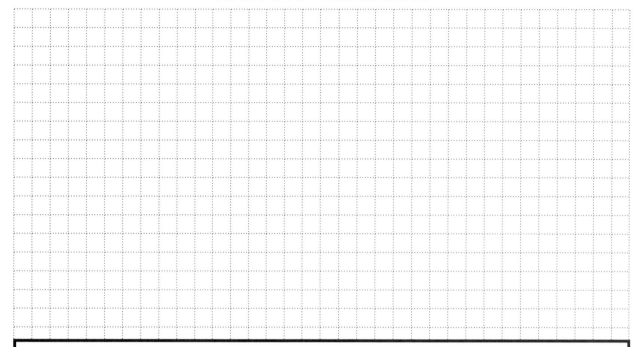

FIGURE A

To draw latitude lines on an inflated round balloon: (1) Inflate a round balloon and tie it closed with tape. DO NOT TIE A KNOT TO SEAL IT. You will need to open it later. (2) Place the balloon down on the table and hold a felt tip pen steady at the balloon's "equator." (3) Slowly rotate the balloon horizontally to draw the equator. (4) Change the position of the pen and draw five more latitude lines above and below the equator. Mark the North and South Poles at the top and bottom of the balloon making a "bull's eye" with the lines you just drew. Each latitude line represents 15º in latitude.

To draw longitude lines on an inflated round balloon: (1) Place the balloon on its side so that its equator is perpendicular to the table. (2) Hold the felt tip pen steady and slowly rotate the ball horizontally to draw a line through the poles around the ball. Label this line the "PRIME MERIDIAN." (3) Roll the balloon over so that both the equator and the prime meridian are at right angles with the table. (4) Draw a second longitude a quarter of the way around the balloon through the poles of the ball. (5) Draw two more lines in between those lines. Each longitude line represents 30º in longitude.

Mark and letter your planet with ten "continents." Then, list the latitude and longitude of each continent to identify the approximate center of each land mass.

	lat.	long.		lat.	long.
A			F		
B			G		
C			H		
D			I		
E			J		

6

MAPPING THE EARTH

Work Date: ____/____/____

LESSON OBJECTIVE

Students will use a compass and a sundial to tell time and direction.

Classroom Activities

On Your Mark!

Before the start of class determine the direction of "magnetic north" using a compass or suspended bar magnet. At the start of class, ask students to point north. Draw Illustration C showing the constellations normally used to locate the North Star (e.g., Polaris): the Big and Little Dipper. The Southern Cross is visible from most any point in the Southern Hemisphere. Explain that daylight navigation requires a magnet able to align itself with the earth's magnetic field. Explain how a compass is used to determine direction. Since the compass always points to magnetic north, north, south, east, and west can be indicated on a piece of paper laid beneath the compass needle. As long as the symbol for north on the paper (e.g., N) is aligned with the needle all directions can be determined. Since the horizon is a circle of 360° we can map landmarks along the horizon in **degrees azimuth**. North is 0° azimuth and South is 180° azimuth. Show students how to use the *Azimuth Indicator* on Journal Sheet #2 to find the location of landmarks along the horizon. Ask students to describe how they would tell the time of day without a wristwatch or digital timing device. Answers will most likely involve the movements of the sun, moon, and stars. A **clock** is a device that measures time by counting the passage of events occurring at regular intervals. Point out that the ancient Egyptians used a "water clock" to tell time. Point out that the drops falling from a leaky faucet fall at a regular rate. The Egyptians used this idea by letting water drip through a small hole at the bottom of a clay pot. They told time by reading the water level markings inside the pot. The first clock designed around the regular swing of a pendulum was invented by **Christiaan Huygens** (b. 1629; d. 1695) in 1657. Early mechanical clocks were driven by pendulums and springs. Modern digital clocks are driven by vibrating crystals sensitive to the pulse of electric current.

> ### ILLUSTRATION C
>
> "Pointer stars" in Big Dipper pointing to the North Star handle in Little Dipper
>
> The Southern Cross

Get Set!

Show students how to construct and read a compass as described in Figure B on Journal Sheet #2. Explain how they can use their hand as a sundial as explained in Figure C on Journal Sheet #2.

Go!

Give students time to construct and use the compass described in Figure B on Journal Sheet #2. Have them record the location of the sun (if you can take them outside) and several other landmarks (either in or out of the classroom). Have them use their "hand sundial" outside.

Materials

ring stand and clamps, bar magnets, string, pens/pencils

EA1 JOURNAL SHEET #2

MAPPING THE EARTH

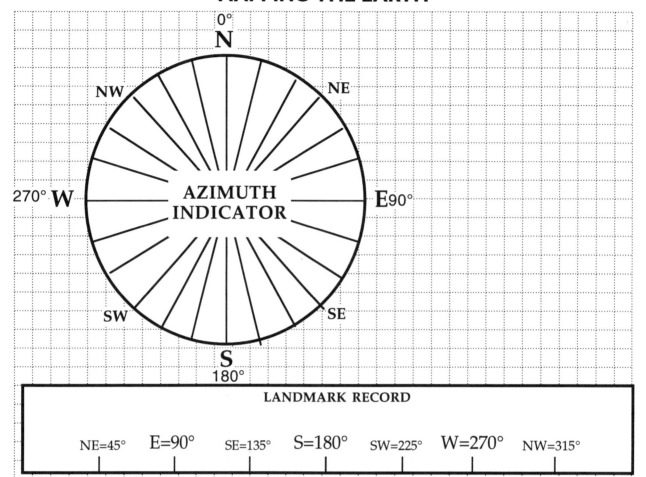

LANDMARK RECORD

NE=45°   E=90°   SE=135°   S=180°   SW=225°   W=270°   NW=315°

FIGURE B

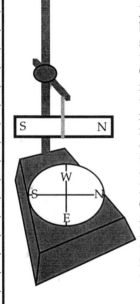

Directions: (1) Suspend a bar magnet from a ring stand as shown. (2) Trace the AZIMUTH INDICATOR above and place it under the magnet so that north (e.g., 0° azimuth) is aligned with magnetic north. (3) Mark or draw the location of the sun, hill or mountaintops, and scattered landmarks along the circumference of the horizon on the LANDMARK RECORD. (4) Note the time. Explain why it is important to note the time.

FIGURE C

Face **west** in A.M.   Face **east** in P.M.

Directions: (1) Use a compass to determine direction. (2) Hold a pencil as shown between your thumb and palm so that the pencil is raised above the palm at about a 45° angle. (3) Face west in the morning and east in the afternoon. (4) Note the shadow cast toward the tip of a finger, finger joint, or side of the palm and read the time as shown.

MAPPING THE EARTH

Work Date: ____/____/____

LESSON OBJECTIVE

Students will use a local map to determine the distance between points of interest.

Classroom Activities

On Your Mark!

Maps are models: scaled representations of actual locations. Give students an example of each of the three types of scales used by mapmakers to show distances on maps. Have students copy each example on Journal Sheet #3. Example #1: A **written** or **verbal scale** states the measurement as a proportion: e.g., "One inch is equal to one mile." Example #2: A **graphic scale** shows a line divided into equal parts. Distances on the map can be compared to the distances between the units on the line. Most freeway and road maps use this type of scale. Example #3: A **fractional scale** states the proportionate relationship between the map and actual distances as a "representative" fraction: e.g., 1 : 125,000. In the latter example, the actual distance is 125,000 times larger than the distance on the map. So, 1 inch on the map would be equal to 125,000 inches ≈ 10,416 feet ≈ 2 miles. Most maps made by the United States Geological Survey use scales of 1:24,000, 1:62,500, or 1:125,000.

Get Set!

Allow students several moments to brainstorm the answers to the questions in the activity described in Figure D on Journal Sheet #3. Spend some time discussing the questions and answers so that students know how to use the fractional scale employed in this activity. Answers: (1) about 6 miles; (2) about 3 miles; (3) absolutely. The length of the park is 1 half inch (≈ 1 mile) and nearly as wide.

Go!

Distribute maps of cities, states, and nations so that students can find the distances between different points of interest. Circulate the room during the activity asking students to tell you the type and meaning of the scale used on their map. Ask them to explain the legend. Have them make a table of distances that lists what they find.

Materials

rulers, maps

EA1 JOURNAL SHEET #3

MAPPING THE EARTH

FIGURE D

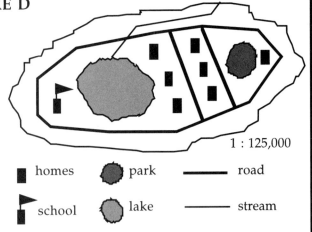

<u>Directions</u>: Use the scale, legend, and a ruler to answer questions #1, #2, and #3. NOTE: There are 5,280 feet in 1 mile.

1. What is the distance run by a jogger who jogs around the island on the road?

2. How far is the shortest bus ride to school from the house by the park?

3. Is the park large enough to hold a football field? Explain your answer.

1 : 125,000

homes   park   ——— road

school   lake   ——— stream

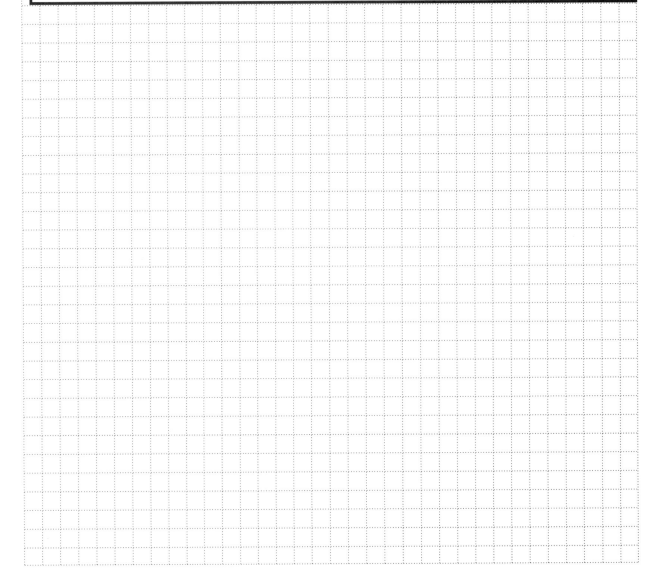

MAPPING THE EARTH

Work Date: ____/____/____

LESSON OBJECTIVE

Students will draw a contour map and construct a corresponding relief model.

Classroom Activities

On Your Mark!

Ask students to consider how they would show the presence of a mountain on a map. This type of information is important as anyone who has ever driven a car over a mountain pass knows. Ask students to reflect on the problems they might encounter if they didn't know they needed to cross a mountain range to get to their destination. **Topographic maps** display this type of information using **contour lines** that connect points of equal elevation. They can be labelled with the appropriate elevation above sea level or colored to give the same information. Contour lines show the "relief" of an area. **Relief** refers to the difference in elevation between the highest and lowest points in a region.

Get Set!

Draw a contour map on the board such as the one in Illustration D. Ask students to answer the following questions: How high is the small mountain? Answer: 400 feet above sea level. Which mountain has the steepest slope: east or west? Answer: the western slope of the eastern mountain; because the contour lines are close together.

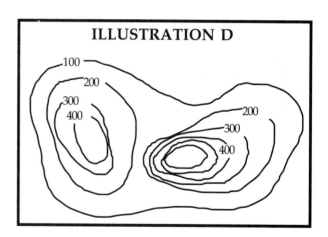

ILLUSTRATION D

Go!

Give students ample time to complete the activity described in Figure E on Journal Sheet #4.

Materials

cardboard, scissors, glue

EA1 JOURNAL SHEET #4

MAPPING THE EARTH

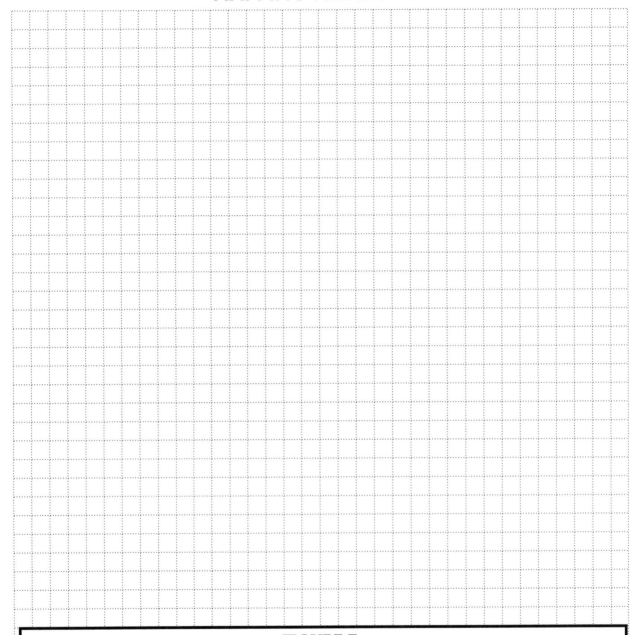

FIGURE E

<u>Directions</u>: (1) Cut out 8-10 patches of cardboard from a large piece of cardboard making each piece slightly smaller than the previous one. Vary the shapes. (2) Place the shapes one at a time on the Journal Sheet and outline them to create a pattern that looks something like the drawing at right. (3) Label each line with an "elevation" of 100 feet, 200 feet, 300 feet, etc. (4) Stack the shapes from the largest to the smallest and glue them together so that they look like your Journal Sheet drawing when viewed from above. (5) Draw a side view of your "mountain" and compare it to the actual model. How would you draw a contour map of a volcano?

EA1 Review Quiz

Directions: Keep your eyes on your own work.
Read all directions and questions carefully.
THINK BEFORE YOU ANSWER!
Watch your spelling, be neat, and do the best you can.

CLASSWORK	(~40): _____
HOMEWORK	(~20): _____
CURRENT EVENT	(~10): _____
TEST	(~30): _____
TOTAL	(~100): _____

(A ≥ 90, B ≥ 80, C ≥ 70, D ≥ 60, F < 60)

LETTER GRADE: _____

TEACHER'S COMMENTS: _____

_____

MAPPING THE EARTH

TRUE–FALSE FILL-IN: If the statement is true, write the word TRUE. If the statement is false, change the underlined word to make the statement true. *15 points*

_____ 1. A map is a <u>picture</u> of the earth's surface.

_____ 2. The astronomer <u>Ptolemy</u> developed the theory that the earth was the center of the universe.

_____ 3. The mapmaker <u>Mercator</u> published the first "flat" map showing plotted points taken from a curved surface.

_____ 4. The <u>North Star</u> is always in a "fixed" position in the sky relative to other stars.

_____ 5. The ancient Greeks noticed that a piece of <u>lodestone</u> hung from a string always pointed toward the North Star.

_____ 6. The astronomer Shen Kua invented a <u>sundial</u> using a magnetized needle, a piece of straw and a bowl of water.

_____ 7. <u>Parallels</u> are drawn vertically on a map with the north direction positioned at the top of the map.

_____ 8. On a globe, all meridians <u>run parallel to</u> the North and South Poles.

_____ 9. Meridians called longitude lines are used to find positions east or west of the <u>equator</u>.

_____ 10. As earth rotates on its axis, the new day begins at <u>midnight</u> at the International Date Line.

_____ 11. <u>Meridians</u> are drawn horizontally on a map, west to east across the map.

_____ 12. A latitude line measures distances north or south of the earth's <u>Prime Meridian</u>.

_____13. The <u>equator</u> is an imaginary latitude line representing 0° latitude.

_____14. The North and South Poles are at <u>90° latitude</u>.

_____15. An earth scientist is called a <u>biologist</u>.

PROBLEM

Directions: Study the contour map shown below and answer questions #16 through #20. *15 points*

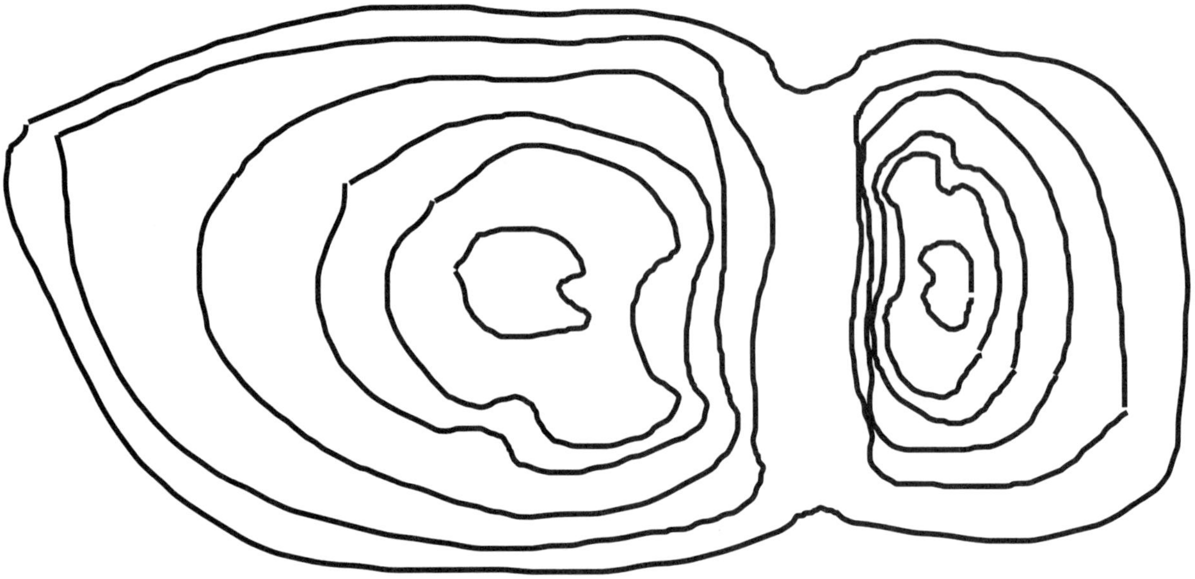

16. Using your pencil or pen "shade in" the valley.

17. Which face of the Western Mountain is steeper? North? South? East? or West? _____

18. Which face of the Eastern Mountain is steeper? North? South? East? or West? _____

19. Which mountain has the steepest face? East or West? _____

20. Color the drawing below that shows the best side view of the map as viewed from the south.

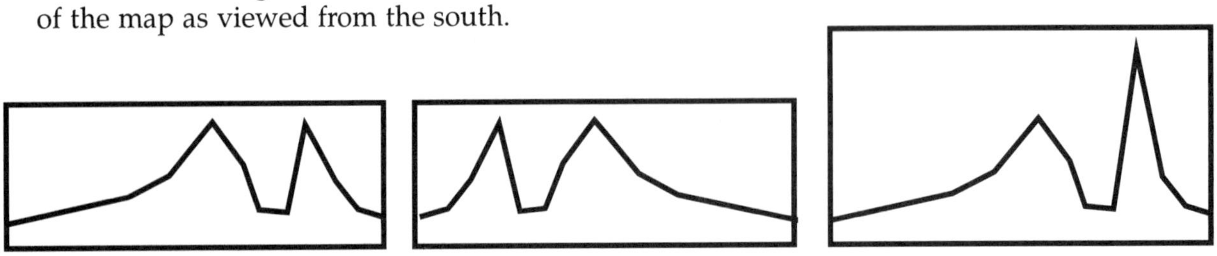

_____ _____ ___/___/___
Student's Signature Parent's Signature Date

14

THE STRUCTURE OF THE EARTH

TEACHER'S CLASSWORK AGENDA AND CONTENT NOTES

Classwork Agenda for the Week

1. Students will show how earthquake waves help us to determine the earth's internal structure.
2. Students will discuss evidence that earth's continents have moved across the planet's surface.
3. Students will show how moving crustal plates build mountains and ocean trenches.
4. Students will show how heat from the earth's interior produces geysers and volcanoes.

Content Notes for Lecture and Discussion

The violent movement of the earth's crust has both intrigued and terrified humankind for thousands of years. History is filled with stories of destruction and death resulting from earthquakes, volcanic eruptions, and the devastating ocean tidal waves that frequently follow them. While ancient people attributed these phenomena to the action of the gods or the accumulation of poisonous gases deep within the interior of the earth, it was not until this century in 1915 that German geophysicist **Alfred Lothar Wegener** (b. 1880; d. 1930) proposed in his treatise *Origin of Continents and Oceans* a comprehensive theory to explain these events. His evidence was largely based on the geographic appearance of the earth's major continents like Africa and South America whose coastlines seem to be part of the same, neatly fitting "global puzzle." Paleontological evidence at the time also supported Wegener's hypothesis by showing clear similarities between the fossilized plant and animal life of each region. Today, Wegener's **theory of plate tectonics** is the guiding theory of **structural geology**.

Geologists of the 18th century confined their investigation to the study of rocks and the formation of **rock strata**. However, the invention of the **pendulum seismograph** in the 1880s by the English seismologist **John Milne** (b. 1850; d. 1913) allowed geologists to study the dynamics of earth's violent upheavals. The study of **seismic waves** using seismometers like the one shown in Illustration A in Lesson #1 has since yielded the most promising clues to the study of the earth's interior. In the late 19th century, the English seismologist **Richard Dixon Oldham** (b. 1858; d. 1936) identified three distinct seismic waves: primary (P) waves; secondary (S) waves; and surface (L) waves. While all of the waves travelled outward from the focus of the geological disturbance each wave had its own distinctive characteristics. P and S waves move through the interior of the earth to arrive at seismic stations at varying locations around the world. P waves, however, travel faster than S waves. L waves are restricted to the surface of the earth unable to penetrate very far into the denser layers of the crust. Oldham knew the velocity of P and S waves and could predict the time it would take for the earthquake waves to travel through the earth and reach the other side of the globe. However, he noticed that at certain distances, thousands of miles from the earthquake focus the waves could not be detected at all. Yet, they seemed to reappear suddenly at more distant seismometers. In addition, the P waves did not arrive at the times predicted based on their calculated velocity. They were delayed. It was as though the waves had been slowed by some obstruction along their journey. In 1906, Oldham decided that something inside the earth was getting in the way of the P waves. He suggested that the earth had a solid **core** that could block or deflect seismic waves. In 1914, the German seismologist **Beno Gutenberg** (b 1889; d. 1960) showed that the outer boundary of the core was located about 1800 miles (≈2900 kilometers) beneath the planet's surface. This boundary is called the **Gutenberg discontinuity**. In 1936, Denmark seismologist **Inge Lehmann** proposed that the earth has an even deeper inner core that bends P waves upward toward the surface. Her hypothesis received enormous support from studies of the seismic waves produced by the nuclear bomb tests of the 1950s. Today, seismologists can determine the densities of the earth's layers using seismic data: crust, 3.0 g/cc; mantle, 4.5 g/cc; outer core, 10.5 g/cc; and inner core, 11.5 g/cc. This makes our planet the densest object in the solar system: even more dense than the sun.

In Lesson #1, students will follow the reasoning of the seismologists mentioned above to show how earthquake waves help us to determine the earth's internal structure.

In Lesson #2, students will discuss evidence listed on their Fact Sheet that earth's continents have moved across the planet's surface.

In Lesson #3, students will show how moving crustal plates build mountains and ocean trenches.

In Lesson #4, students will show how heat from the earth's interior produces geysers and volcanoes.

ANSWERS TO THE HOMEWORK PROBLEMS

Student maps will vary but should indicate that the student has a grasp of latitude and longitude. They should also recognize that the youngest volcanic islands are located east of the oldest volcanic islands. This should lead them to the conclusion that the planet has an active mantle and moving crust. They should mention that stationary "hot spots" in the planet's mantle create the islands one after another as the crust moves across the surface of the planet. They should be able to calculate that the crustal plate containing these islands is moving at about 10° longitude every 5 million years.

ANSWERS TO THE END-OF-THE-WEEK REVIEW QUIZ

1. true	6. true	11. A
2. different	7. Wegener	12. B
3. tectonics, structural geology	8. true	13. C
4. true	9. faster	14. E
5. true	10. true	15. D

PROBLEM

"trench" points to D
"ridge" points to A
"plate" points to C
"densest iron" points to E
"fluid iron" points to B

EA2 FACT SHEET

THE STRUCTURE OF THE EARTH

CLASSWORK AGENDA FOR THE WEEK

(1) Show how earthquake waves help us to determine the earth's internal structure.
(2) Discuss evidence that earth's continents have moved across the planet's surface.
(3) Show how moving crustal plates build mountains and ocean trenches.
(4) Show how heat from the earth's interior produces geysers and volcanoes.

An **earthquake** shakes the ground causing mechanical energy to move across the planet's surface and through its interior. Waves of mechanical energy called **seismic waves** are recorded on a device called a **seismometer**. Seismographic recordings on a seismometer show that the waves move through the planet at different speeds. This fact allows scientists to locate the places where earthquakes start and to discover much about the internal structure of our planet. The study of the earth's structure is called **structural geology** or **tectonics**. In the past century, structural geologists have discovered that the earth is divided into three major structures: a *crust*, a *mantle*, and a *core*. The **crust** is the thin layer of rock that surrounds the planet like the thin peel of an apple. It varies in thickness from about 3 to 40 miles (≈ 5 to 65 kilometers). The denser **mantle** is made of molten iron and nickel and plunges to a depth of about 1,800 miles (≈ 2,900 kilometers). The earth's center or **core** is made of very dense iron and nickel extending about another 2,160 miles (≈ 3,480 km) to the center of the planet.

Many theories have been proposed to explain the cause of earthquakes. Scientists of the 16th and 17th centuries thought that earthquakes were the result of accumulating poisonous gases and explosive chemicals. Today, the overwhelming evidence points to another theory. This new theory is called the *theory of plate tectonics*. The **theory of plate tectonics** is a theory of "continental drift." It was first proposed by German geophysicist **Alfred Lothar Wegener** (b. 1880; d. 1930). Wegener suggested that the earth's crust is broken into large chunks, or **plates**. He proposed that currents in the molten magma of the mantle force the plates to move across the earth's surface. Today, geologists can measure the rate at which the continental plates are moving. On average, the continents travel several inches (≈ 2-4 centimeters) every year. But given enough time (e.g., millions of years) they will move a considerable distance. The theory of plate tectonics is supported by several pieces of evidence: (1) the movement of the plates can be measured, (2) the edges of distant continents resemble puzzle pieces that can be assembled into a large "mother continent," (3) identical plant and animal fossils found on distant continents are the remains of organisms that must have evolved in exactly the same place, and (4) the earth's major earthquake zones are located in regions of maximum earthquake activity that mark out the edges of the continental plates.

When magma in the molten mantle flows, the crustal plates move. Stress is created in the rocks which crack and buckle under the strain. The breaking, tilting, and folding of rocks is called **deformation**. **Compression** squeezes rocks together. **Tension** pulls rocks apart. **Shearing** pushes rocks in opposite directions causing them to twist and tear. A break or crack along which rocks move is called a fault. **Mountains** are formed by "uplifting" sections of crust called **blocks**. **Valleys** form when blocks of land "fall" between two separating blocks of land. A "bend" in a rock is called a **fold**. Folds can be identified by examining the "twisted" contour of rock layers called **strata**.

Magma that reaches the Earth's surface is called **lava**. The place where magma reaches the Earth's surface is called a **volcano**. Not all volcanoes are found on land. Most volcanoes are located in the depths of the ocean where the Earth's crustal plates are spreading apart. The opening where lava erupts to the surface is called a **vent**. The funnel-shaped pit, or depression, at the top of a volcano is called a **volcano cone** or **crater**. When the walls of the crater collapse a **caldera** forms that can fill with water to produce a lake. Studying the minerals in lava gives scientists clues to the composition of the Earth's interior.

Homework Directions

Imagine you are a space explorer employed to examine the structure of an alien world. Upon arriving on the newly discovered planet, you record the locations of a group of oceanic islands according to their latitude and longitude. Since you are the first human explorer to visit the planet you can pick the location of the planet's Prime Meridian. Use a piece of graph paper to plot the location of the islands listed in Table A. In addition to plotting the location of the islands, you also test the rocks on each island to determine their age. You record that information as well and write the age of each island on your graph. You are ordered to write a paragraph report to your superiors at Exploration Command Center. The report must describe the geologic structure of this new planet by answering the following questions: (1) Does the planet have a moving crust like that of earth? (2) If it does, how fast do the pieces of the crust move? You must thoroughly explain your reasons for drawing your conclusions.

TABLE A		
island	age	location
A	1,000 years	5° N. lat; 10° W. long.
B	1,000,000 years	10° N. lat; 20° W. long.
C	5,000,000 years	5° N. lat; 30° W. long.
D	10,000,000 years	0° lat; 40° W. long.
E	15,000,000 years	5° S. lat; 50° W. long.

Assignment due: _____

_____   _____   ____/____/____
Student's Signature   Parent's Signature   Date

THE STRUCTURE OF THE EARTH

Work Date: ____/____/____

LESSON OBJECTIVE

Students will show how earthquake waves help us to determine the earth's internal structure.

Classroom Activities

On Your Mark!

Draw the **seismometer** in Illustration A on the board and have students copy your drawing on Journal Sheet #1. Be sure they identify the **primary (P)**, and **secondary (S) seismic waves** discovered by English geologist **Richard Dixon Oldham** (b. 1858; d. 1936) in 1897. Explain that geologists use the fact that P and S waves move at different speeds to find the **focus** and **epicenter** of an earthquake. Ask students to imagine the following: They are riding on their bicycles followed by a little brother or sister who cannot go as fast as they can. If they move at 20 feet per second and their smaller sibling rides at 10 feet per second, how far behind would their brother or sister be after one second? Two seconds? Three? Answer: 10 feet, 20 feet, and 30 feet. Their sibling will fall farther and farther behind the farther they ride. Explain that P waves travel faster than S waves. So, geologists can calculate where the waves began by measuring the distance between waves on different seismometers. A station by the quake will record P and S waves close together. A seismometer far from the quake will record the waves far apart. A minimum of three seismograph stations are used to "triangulate" the epicenter of an earthquake. Refer to Illustration B.

ILLUSTRATION A

The seismometer is anchored to the foundation of the building so that the pen suspended by springs can vibrate up and down and draw lines on a rotating drum covered with graph paper.

primary wave

secondary wave

ILLUSTRATION B

Seismic stations at cities X, Y, and Z register quakes. By measuring how long it took the S wave to follow the P wave at each seismometer, geologists can calculate how far away the quake was from their station (e.g., anywhere on the circle surrounding their station). Three stations are used to find the exact location of the quake. The location of the quakes is where the three circles cross at the dark dot.

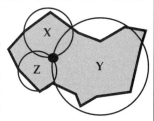

Get Set!

Use the Teacher's Agenda and Content Notes to explain Oldham's reasoning for postulating the existence of earth's core. Use the circle on Journal Sheet #1 to help students to visualize the **crust**, **mantle**, and **core** described in their Fact Sheet. Have them write the names of these layers of the earth.

Go!

Give students ample time to complete the activity described in Figure A on Journal Sheet #1.

Materials

paper, flashlights, pencils, rulers, tape

EA2 JOURNAL SHEET #1

THE STRUCTURE OF THE EARTH

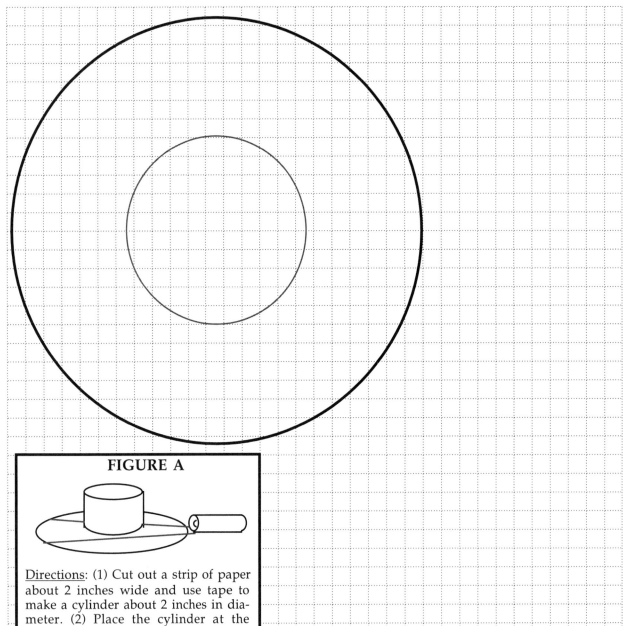

FIGURE A

Directions: (1) Cut out a strip of paper about 2 inches wide and use tape to make a cylinder about 2 inches in diameter. (2) Place the cylinder at the center of the circle above. (3) Turn on the flashlight and place it on the edge of the circle representing the earth's crust. The light from the flashlight simulates the "waves" moving through the earth following an earthquake. (4) Use pencil to shade in the shadow created by the paper cylinder. (5) Change the position of the flashlight to simulate the "shadow zones" created by earthquakes occurring at five more places in the crust.

THE STRUCTURE OF THE EARTH

Work Date: ____/____/____

LESSON OBJECTIVE

Students will discuss evidence that the earth's continents have moved across the planet's surface.

Classroom Activities

On Your Mark!

Prepare a handout displaying disconnected "continents" that can be puzzled back together to form a "supercontinent" named after your school. Make sure the edges of each continent have the same "fossils" (e.g., represented by letters or symbols) to match those on adjoining continents (cf., activity described in Figure B).

Display a map of the earth and point out how the east coast of South America and the west coast of Africa seem to be shaped like the edges of neatly fitting puzzle pieces. Point out that in 1915 the German geophysicist **Alfred Lothar Wegener** (b. 1880; d. 1930) suggested in his book, *Origin of Continents and Oceans*, that the two continents were once joined. This suggestion is supported by the fact that identical plant and animal fossils in identical layers of earth are found on these two coasts and nowhere else on the planet. Explain that while "similar" species may differ in terms of their time and place of evolutionary origin, members of an

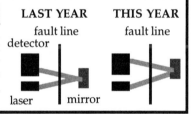

ILLUSTRATION C

A laser beam is aimed at a mirror across a faultline and the angle of reflection is registered on a detector. The results of this type of experiment done at many geological observatories around the world indicate that landmasses across faultlines move relative to one another at several centimeters per year.

LAST YEAR
fault line
detector

THIS YEAR
fault line

laser   mirror

identical species must have first appeared on earth at the same time and place. Mention as well that scientists can measure the motion of structures placed on opposite sides of major earthquake faults like the San Andreas Fault in California. Draw Illustration C to explain this and have students copy your drawing on Journal Sheet #2. Explain that Wegener's "theory of continental drift" is now called the **Theory of Plate Tectonics**. Geologists believe that energetic currents in the molten magma of the earth's mantle push the cracked sections of the earth's crust, called **tectonic plates**, across the planet's surface. Earthquake and volcanic activity occur most often at places where the plates collide and slide against one another.

Get Set!

Point out that the crust averages 40 miles thick compared to 4,000 miles to the earth center. The crust is therefore only about 1% the radius of the earth. Perform the following demonstration to illustrate the principle of continental drift: (1) Place a tin pie plate on a hot plate. (2) Pour several millimeters of corn oil into the plate. (3) Cut small sections of cardboard and fit the pieces together on the surface of the oil. (4) Turn on the hot plate to a low setting and allow the corn oil to warm. (5) Return to the set up every few minutes so that students can observe how the cardboard "plates" have shifted. Explain how this model suggests the way in which the earth's crustal plates move across our planet's surface.

Go!

Assist students in performing the activity described in Figure B on Journal Sheet #2.

Materials

pie plates, corn oil, cardboard, hot plates, construction paper, glue, scissors

EA2 JOURNAL SHEET #2

THE STRUCTURE OF THE EARTH

FIGURE B

Directions: (1) Cut out the individual pieces of "continent" including the "fossil shapes" given to you by your instructor. (2) Glue the pieces together on a separate sheet of construction paper so that all of the separate landmasses make one giant "supercontinent." Be sure to match identical fossils and coastlines. (3) Write several sentences to explain how each individual piece must have drifted over the eons in order to arrive in the positions shown here. (4) Consider the way these landmasses have shifted about. Would you say that this planet had a peaceful or violent geological history?

THE STRUCTURE OF THE EARTH

Work Date: ____/____/____

LESSON OBJECTIVE

Students will show how moving crustal plates build mountains and ocean trenches.

Classroom Activities

On Your Mark!

Refer students to the map of the world on Journal Sheet #3 and have them identify the world's continents. Point out the thick-lined borders of the **continental plates** where the earth's major crustal plates **converge** and **diverge**. Review the forces discussed in Lesson #2 that cause the earth's plates to move across the planet's surface. See if students can identify the smaller Arabian, Phillipine, Nazca (west coast of South America), Cocos (west coast of Central America), and Caribbean plates also indicated on the map. Draw Illustration D and describe the mountain-building events that occur at Pacific convergent zones. Identify major **oceanic trenches** in the Pacific Ocean (such as the Marianas Trench off the eastern border of the Phillipine Plate) where plates collide: one plate being subducted back into the mantle under another. Point out that new crust is continually being created at divergent zones called **rift valleys** where the sea floor is spreading apart. Have students note all of these features on the map.

> ### ILLUSTRATION D: PLATES IN COLLISION
> Plates are pushed apart at diverging zones by upwelling magma from the mantle. The magma cools in the depths of the ocean to form new crust. On the other side of the world at converging zones the enlarging plates collide, the heavier rock sinking back into the mantle. At "hot spots" around the globe magma can burst through the crust in the middle of a plate creating volcanic mid-oceanic islands like the Hawaiian Islands.
>
>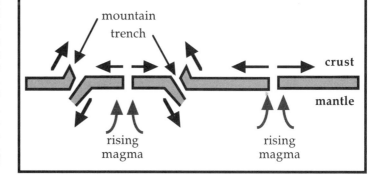

Get Set!

Explain to students that rock layers in the crust "deform" when they are under stress. Where plates press together slowly rock layers **fold**. Where plates crash into one another more quickly they may **fracture** or **fault**. The density of the rock layers and the direction at which they collide determines the variety of topographical features that form: **fold mountains**, **normal faults**, **reverse faults**, and **lateral faults**.

Go!

Assist students in performing the activities described in Figure C on Journal Sheet #3 in order to help them distinguish between the different types of faults.

Materials

different colors of modelling clay

Name: _____   Period: _____   Date: ___/___/___

EA2 JOURNAL SHEET #3

THE STRUCTURE OF THE EARTH

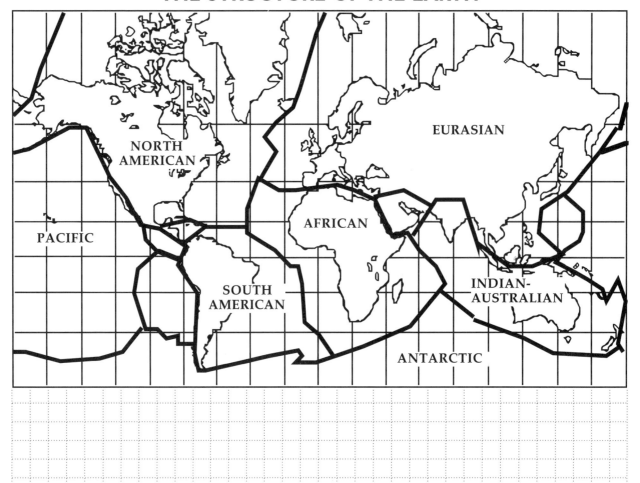

FIGURE C

<u>Directions</u>: (1) Use three different colors of modelling clay to make 6 squares each about 3 inches wide (e.g., 2 squares per color). (2) Layer 3 squares of different color to make 2 "crustal plates." (3) Press the plates together at varying speeds to see if you can produce the 4 types of topographical features shown below. (4) Draw the formations you create.

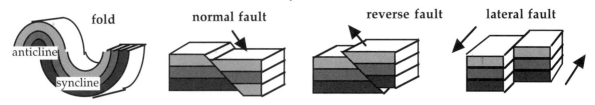

THE STRUCTURE OF THE EARTH

Work Date: ___/___/___

LESSON OBJECTIVE

Students will show how heat from the earth's interior produces geysers and volcanoes.

Classroom Activities

On Your Mark!

Review the forces that cause the earth's crustal plates to move. Remind students that currents in the molten magma of the mantle push the plates apart or into collision. Any place where molten magma rises to the earth's surface is called a **vent**. **Volcanoes** form where magma cools and hardens on the surface. Most of our planet's volcanoes are located under the ocean at mid-ocean ridges. Others are found at places where the plates collide. Some, like the Hawaiian Islands, are found in the middle of a plate over a portion of mantle called a **hot spot**. Volcanoes erupt when the pressure caused by magma and expanding gases inside the volcano becomes greater than the strength of the volcano's hardened walls. **Geysers** form in regions that are volcanically active. A geyser is a hot spring where water over a vent is heated to extreme temperatures and turned into steam. The steam is forced out of the cracks in the earth where it gushes out in dramatic bursts. The country of Iceland has about 30 active geysers because it lies along the path of the Mid-Atlantic Ridge. Yellowstone National Park in California, a volcanically active region that last erupted about 600,000 years ago, is noted for more than 200 geysers. The park is located where the Pacific and North American plates slide past one another.

Get Set!

Have students discuss the drawings in Figure D on Journal Sheet #4 and answer the questions together. Discuss their answers as a class to make sure that everyone understands how volcanoes form. <u>Answers</u>: (1) four; (2) the third because the layer of ash is the thickest; (3) Yes, because the volcano erupts every 25,000 years and the last eruption was 25,000 years ago.

Go!

After reviewing appropriate safety procedures, give students ample time to complete the activity described in Figure E on Journal Sheet #4. They will observe the geyser erupting in bursts as pressure builds to a maximum inside the funnel before each spurt of hot steam.

Materials

hot plates, large beakers, glass funnels, goggles, heat resistant gloves and aprons

EA2 JOURNAL SHEET #4

THE STRUCTURE OF THE EARTH

FIGURE D

100,000 years ago   75,000 years ago   50,000 years ago   25,000 years ago

1. How many times has this volcano erupted? Explain.

    _____

    _____

2. Which eruption was probably the most violent or lasted the longest? Why?

    _____

    _____

3. Would you feel confident in predicting another eruption within the next few thousand years? Explain your answer.

    _____

    _____

FIGURE E

Directions: (1) Fill a large 1,000 ml beaker with about 250 ml of water. (2) Invert a glass funnel into the water as shown. (3) Turn on the hot plate to a medium setting. (4) Record your observations after 1, 2, 3, 4, 5, 6, 7, 8, 9, and 10 minutes. Does the "geyser" erupt in a continuous flow or does it erupt in bursts? Explain your answer.

GENERAL SAFETY PRECAUTIONS

Wear goggles, heat resistant gloves, and an apron. STEAM CAN CAUSE SERIOUS BURNS. Be sure you are familiar with the proper way to use the hot plate. Discard and clean up as instructed by your teacher only after the apparatus has cooled.

EA2 Review Quiz

Directions: Keep your eyes on your own work.
Read all directions and questions carefully.
THINK BEFORE YOU ANSWER!
Watch your spelling, be neat, and do the best you can.

CLASSWORK	(~40): _____
HOMEWORK	(~20): _____
CURRENT EVENT	(~10): _____
TEST	(~30): _____
TOTAL	(~100): _____

(A ≥ 90, B ≥ 80, C ≥ 70, D ≥ 60, F < 60)

LETTER GRADE: _____

TEACHER'S COMMENTS: _____
_____

THE STRUCTURE OF THE EARTH

TRUE–FALSE FILL-IN: If the statement is true, write the word TRUE. If the statement is false, change the underlined word to make the statement true. *20 points*

_____ 1. Earthquake waves are recorded on a device called a <u>seismometer</u>.

_____ 2. Seismographic recordings show that the waves move through the planet at <u>the same</u> speeds.

_____ 3. The study of the earth's structure is called <u>technology</u>.

_____ 4. The earth's <u>crust</u> is about 3 to 40 miles (≈ 5 to 65 kilometers).

_____ 5. The earth's <u>mantle</u> is made of molten iron and nickel and plunges to a depth of about 1,800 miles (≈ 2,900 kilometers).

_____ 6. The earth's <u>core</u> at the center of the planet is made of very dense iron and nickel.

_____ 7. The theory of plate tectonics was first proposed by geophysicist <u>Sir Isaac Newton</u>.

_____ 8. The breaking, tilting, and folding of rocks is called <u>deformation</u>

_____ 9. P waves are <u>slower</u> than S waves.

_____ 10. Magma that reaches the Earth's surface is called <u>lava</u>.

MATCHING: Choose the letter of the word or phrase at right that best defines the vocabulary term at left. *5 points*

_____ 11. compression   (A)   rocks pushed together

_____ 12. tension   (B)   rocks pulled apart

_____ 13. shearing   (C)   rocks sliding against one another

_____ 14. faulting   (D)   rock layers bent slowly out of shape

_____ 15. folding   (E)   rock layers quickly broken

PROBLEM

Directions: Identify the zones indicated by the letters A, B, C, D, and E by drawing an arrow from one of the vocabulary words listed to the correct letter. *5 points*

trench   ridge   plate   densest iron   fluid iron

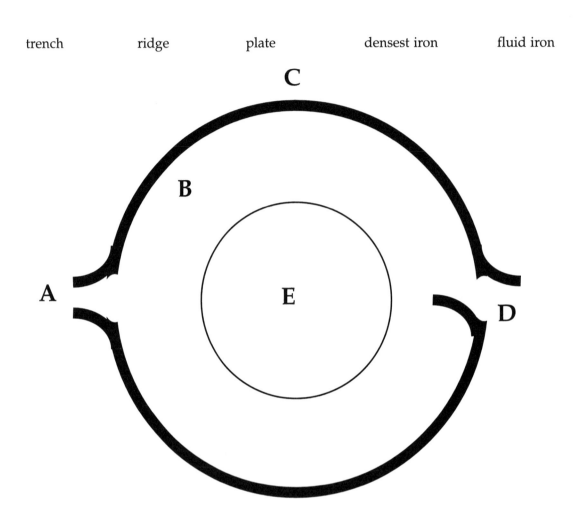

_____   _____   ___/___/___
        Student's Signature   Parent's Signature   Date

28

CHANGING LAND

Teacher's Classwork Agenda and Content Notes

Classwork Agenda for the Week

1. Students will identify the major forces that change the surface of the earth's crust.
2. Students will examine the effects of wind and water erosion.
3. Students will examine the effects of mechanical and chemical weathering.
4. Students will discuss how glaciers have changed the face of continents.

Content Notes for Lecture and Discussion

The predominant view at the end of the 17th century held that the earth was created in a state of perfection which was followed by a process of decay. European theologists attributed this decaying process to the Fall described in the Bible, a view analogous to the obvious notion that decay resulting from age is the fate of all God's creations; so, why not earth? This view changed in the 18th century as philosophers embraced the ideas of the Naturalists who viewed earth as a complex but durable machine. The Naturalists held that all landforms, the atmosphere, and oceans of the globe were parts of an interlocking mechanism that could never run down. The Scottish geologist **James Hutton** (b. 1726; d. 1797), known as the "father of geology," suggested that the earth was incredibly old—a view harshly criticized by the Christian Church. Hutton proposed that the land undergoes continual transformation by much the same processes as those which occurred in times past. Wind blows and rivers run as they always have, eroding and compacting soil and sediment in a continual process of change. Hutton's notions were later extended by the Scottish geologist **Charles Lyell** (b. 1797; d. 1875) and remain the basis for much of the work done by modern geologists. In his book *Principles of Geology* published in 1830, Lyell deduced the age of the earth to be approximately 240,000,000 years old and gave a description of the earth's geological history. He based his conclusions on estimated rates of sedimentation; and although he was completely wrong—more than 4 billion years off—the logic of his argument served to promote and popularize his position. Lyell's book greatly influenced **Charles Darwin** (b. 1809; d. 1882) who formulated **the theory of evolution**.

The American physical geographer **William Morris Davis** (b. 1850; d. 1934) developed one of the first organizing principles of geology: the erosion cycle. And, while the erosion cycle is not entirely accurate in its every detail, it still served as the dominant model of geomorphology for nearly half a century. According to Davis's model, the cycle begins with the uplifting of the land by earthquake or volcanic action; mountains build and erode until they reach a level featureless plain.

In Lesson #1, students will identify the major forces that change the surface of the earth's crust. In the course of lecture, it should be pointed out that **erosion** and **sedimentation** are complementary geological processes. Erosion wears down rocks, removing them from one area and transporting them to another. Sedimentation serves to anchor material in place creating unique strata whose features can be used to interpret the geological history of a region.

In Lesson #2, students will examine the effects of wind and water erosion and learn that gravity is a forceful component of the erosion process. It can be noted that wind erosion occurs mostly on beaches and in deserts. In these places there is little or no groundcover or vegetation to hold the landscape together. Mild breezes can pick up finely grained rock (e.g., grains approximating 1–2 mm in diameter) by a process known as **deflation**. Strong winds can hurl much larger rocks and cause finely grained sand to abrade, erode, and polish rock surfaces. Point out that the same erosion and sedimentation processes probably occur on other planets like Mars. By carefully analyzing photographs returned from distant planets, geologists are trying to piece together the history of these other worlds.

In Lesson #3, students will examine the effects of **mechanical** and **chemical weathering** and distinguish between the physical (e.g., temperature changes) and chemical (e.g., oxidation, etc.) processes that alter the features of the land.

In Lesson #4, students will discuss how glaciers have changed the face of continents.

ANSWERS TO THE HOMEWORK PROBLEMS

Students will observe that ice takes up more space than liquid water. The clay plugs will be pushed out the ends of the straw that was placed in the freezer overnight. This is, of course, an example of mechanical weathering. In Lesson #3, you will show students the soda cans that you filled with water and placed in the freezer overnight. The frozen can will certainly bulge and may possibly split as the water freezes and expands. Caution students against performing this demonstration as the split metal edges of soda cans are very sharp!

ANSWERS TO THE END-OF-THE-WEEK REVIEW QUIZ

1. Geologists
2. true
3. true
4. true
5. Gravity
6. can
7. true
8. glacier
9. mechanical
10. chemical
11. composition of the rock
12. time the rock is exposed
13. surface area of exposed rock
14. erosion
15. weathering
16. sediment
17. weathering
18. gravity
19. wind
20. glaciers

EA3 FACT SHEET
CHANGING LAND

CLASSWORK AGENDA FOR THE WEEK

(1)   Identify the major forces that change the surface of the earth's crust.
(2)   Examine the effects of wind and water erosion.
(3)   Examine the effects of mechanical and chemical weathering.
(4)   Discuss how glaciers have changed the face of the continents.

Look around your neighborhood and think of the ways it has changed since you were a small child. Have the trees grown larger? Has there been any new construction that has replaced a woodland, marsh, or open field? It is probably obvious to you that living things, plants, and animals including people, have the power to change the surface of land features. However, there are additional forces that also change the surface of the land. Geologists identify two major processes that contribute to the wearing down of the land: erosion and weathering.

Erosion is the process by which beaten rock and soil particles are moved from one place and deposited in another. Loose rocks and soil are called **sediment**. Sediment is usually deposited in layers that geologists call **strata**. If you have ever seen pictures of the Grand Canyon, or seen a hillside that was carved away to make room for a highway, you know what strata look like. There are four main causes of erosion: gravity, moving water, wind, and glaciers. **Gravity** pulls on matter at the surface of the earth causing the downhill movement of sediment. This can occur quickly in a landslide or little by little. **Moving water** in ocean waves, streams and rivers, can carve out the landscape over short or long periods of time. A tidal wave or flood can drastically change the surface of the land in a tragically short span of time. It took several million years for the Colorado River to dig the Grand Canyon, washing away sediment at the rate of 500,000 tons per day. The force of the wind can also lift and carry away small or large amounts of sand and sediment. In addition, quickly moving sand, like rapidly moving water, can chisel and reshape both human-made and natural structures as though they were rubbed with sandpaper. A **glacier** is a large mass of moving ice and snow. The continent of Antarctica is presently covered by a glacier. A glacier is formed by the build up of ice and snow during a cold spell that can last thousands of years. Snow accumulates winter after winter during a time when the summers are not warm enough to melt the winter ice. During warm periods, glaciers melt and leave behind rocks and other debris. The Great Lakes on the border of the United States and Canada were carved out by a receding glacier.

Weathering can be either mechanical or chemical. **Mechanical weathering** causes rocks to break into very small pieces without changing the chemical composition of the rocks themselves. Tree roots and temperature changes are just two kinds of mechanical weathering that can cause rocks and cement to crack. During **chemical weathering**, however, the minerals in rocks are changed to form new chemical substances. Air pollution which contains many kinds of acids can burn and dissolve both natural and manmade structures. Many famous monuments, like the Statue of Liberty and the images of Presidents Washington, Jefferson, Roosevelt, and Lincoln engraved into Mount Rushmore, become the victims of chemical weathering. A number of factors can influence the time it takes for land features to become weathered. These factors are as follows: (1) the composition of the rock, (2) the amount of time the rock is exposed to the forces of erosion, and (3) the surface area of the rock that remains exposed. Geologists can tell a great deal about the history of a particular region by studying layers of sediment and other existing landforms in an area. Knowing how erosion and weathering take place today provides clues to what the land might have looked like in the past.

Homework Directions

Perform the following experiment. (1) Fill two paper or plastic straws with water so there are no air bubbles present in the straws. (2) Cap both ends of each straw with clay so that the water cannot escape. (3) Place one straw in the freezer overnight and leave the other on a plate on the kitchen counter for the same period. (4) Draw and record what you observe the following day.

Write a short paragraph that answers the following questions: (a) What happened? (b) Is this experiment an example of erosion? Chemical weathering? Mechanical weathering? Explain your answer.

Assignment due: _____

_____
Student's Signature   Parent's Signature   ____/____/____
                                                                    Date

CHANGING LAND

Work Date: ____/____/____

LESSON OBJECTIVE

Students will identify the major forces that change the surface of the earth's crust.

Classroom Activities

On Your Mark!

Before class compile 40–50 pictures (e.g., flooded towns, river rapids, tornadoes, waves against a beach, stained old statues, etc.) from old magazines and newspapers (e.g., articles and advertisements) that depict landforms in the process of being changed by erosion and weathering. Use a magic marker to number or letter the pictures. At the start of class, review the concept of force. Remind students that <u>a force is a push or pull on an object that results in movement</u>. Have them copy that definition on Journal Sheet #1. Give them five minutes to brainstorm and record in a lefthand column on Journal Sheet #1 any forces they can imagine that result in the transformation of landforms.

Get Set!

At the end of the brainstorming session draw the graphic organizer in Illustration A on the board, have students copy your illustration on Journal Sheet #1, and define each of the terms as they appear in the Fact Sheet. Use the information in the Teacher's Classwork Agenda and Content Notes to give students an historical perspective about the causes of landform changes from the viewpoints of the great geologists like **James Hutton** (b. 1726; d. 1797).

Go!

Distribute the prepared pictures to class groups so that each student has the opportunity to identify the forces of erosion and weathering exemplified in at least 4–6 samples. Instruct them to record the number or letter code of the picture in the appropriate space on the graphic organizer and write a sentence describing the

forces in action in each sample. Circulate the room during the activity and ask students to explain their choice of the erosion/weathering forces involved.

ILLUSTRATION A

EROSION
- gravity
- moving water
- wind
- glaciers

WEATHERING
- mechanical
- chemical

Materials

old magazines, newspapers, scissors, magic marker

Name: _____   Period: _____   Date: ____/____/____

EA3 JOURNAL SHEET #1

CHANGING LAND

CHANGING LAND

Work Date: ____/____/____

LESSON OBJECTIVE

Students will examine the effects of wind and water erosion.

Classroom Activities

On Your Mark!

Before class layer the bottoms of trays (e.g., cardboard soda can boxes) with waxed paper and 2–3 centimeters of fine grained soil or sand so that the soil or sand covers nearly three quarters the length of the box. Use an additional strip of cardboard to separate the box into two halves lengthwise so that students can study the effects of "wind" on one side and "moving water" on the other (cf. Figure A on Journal Sheet #2). Review the definition of erosion as the movement of sediment from one location to another. Explain that wind erodes by **deflation** and **abrasion**, lifting sand and dust particles into the air and rubbing them across the surface of rocks. Sand dunes are created by the continual movement of wind across the rippled surface of the desert. Point out that gravity plays an important role in the process of erosion. Explain that while wind is created by the differential heating of the earth's surface (e.g., warm air rises and cold air moves in to take its place) water always flows downhill under the influence of gravity.

Get Set!

Assist students in setting up the activity described in Figure A on Journal Sheet #2. Go over the procedure and ask them to draw what they think will happen in the Prediction Drawings when the landscape is subjected to light and heavy breezes or slow and fast moving water.

Go!

Give students ample time to complete the activity. Circulate around the room and point out that wind has a "rippling effect" that tends to create small dunes. Running water produces **runoff**: a major product and cause of further erosion. Show them the network drainage system of **rills** and **gullies** that erode regions of land and the deposits of soil at the far end of the box that create **alluvial fans** and **deltas** at the ends of large rivers.

Materials

cardboard soda can boxes, string, scissors, ringstand and clamps, funnel, 250 ml beakers, water, straws, fine grained soil or sand

EA3 JOURNAL SHEET #2

CHANGING LAND

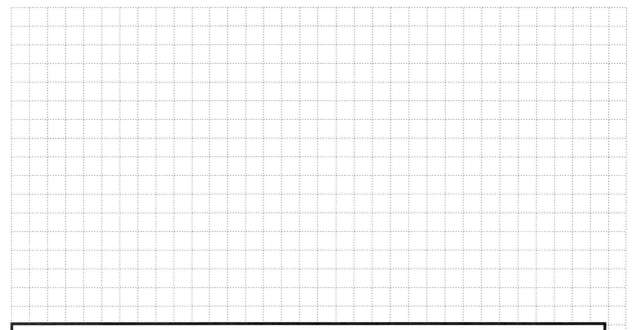

FIGURE A

Directions: (1) Attach the sandfilled tray to the ringstand with a clamp so that one end of the tray is raised at a shallow incline (e.g., 5 cm above the table surface). (2) Clamp the funnel in position on the water RUNOFF side of the tray. (3) Draw your predictions for the effects that will be produced by light and heavy breezes you will be creating by blowing softly and with force through a straw over the sand. (4) Draw your predictions for the effects that will be produced by pouring 50 ml, then 250 ml, water through the funnel. (5) Test your predictions one at a time drawing your results after each test. To make an accurate drawing, you can use string to divide the box into 16 "zones" as shown and draw the effects of erosion in each zone. (6) Write a paragraph that summarizes what you observed. How do your observations compare with your predictions?

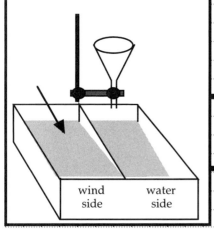

PREDICTION FOR SLOW RUNOFF   PREDICTION FOR HEAVY RUNOFF

PREDICTION FOR LIGHT BREEZE   PREDICTION FOR HEAVY BREEZE

EFFECT OF SLOW RUNOFF   EFFECT OF HEAVY RUNOFF

EFFECT OF LIGHT BREEZE   EFFECT OF HEAVY BREEZE

wind side   water side

EA3 Lesson #3

CHANGING LAND

Work Date: ____/____/____

LESSON OBJECTIVE

Students will examine the effects of mechanical and chemical weathering.

Classroom Activities

On Your Mark!

Prepare for this demonstration at least 24 hours before you present it: (1) Fill two empty soda cans with water up to the rim. (2) Put them in the freezer overnight and allow the water to expand. Begin class discussion by asking students if it is possible for rocks and landscape to change without being "eroded." That is, can rocks change if they are not moved? Show the class the frozen cans filled with ice. Ask students to explain their observations. Explain that water is one of the few liquids that expands when it freezes. Most other liquids contract when they cool. Ask: What would have happened if this container were made of glass? Answer: It would probably have shattered. This demonstration achieves the same results as that which students will obtain in their *Homework* experiment. Point out that a rock cracking as the result of changes in temperature is an example of **mechanical weathering**. Define mechanical weathering as a physical breakdown in rocks that does not change the chemical composition of the material. Ask students if rocks that contain iron can change without erosion. Of course, any rocks that contain iron can rust with exposure to oxygen. Explain that rust is a compound made of iron and oxygen (e.g., Fe_2O_3). Rusting is an example of **chemical weathering** because the iron is no longer pure iron. It is iron oxide.

Get Set!

Perform the following quick demonstration: (1) Put a tablespoon of baking soda into a beaker. (2) Add a tablespoon of vinegar. The bubbles produced are **carbon dioxide** bubbles. Ask students to comment: What would happen if you poured vinegar, a weak household acid, onto a rock that contained baking soda, a compound containing carbonate (e.g., CO_3). Answer: Carbon dioxide bubbles would form on the surface of the rock just as they did in the beaker. Ask: Does the atmosphere contain acid? *Comment:* Ever hear of acid rain? List the three main types of chemical weathering on the board and have students copy them on Journal Sheet #3: **hydrolysis**—the addition of moisture to substances forms hydrated oxides and silicates (e.g., silicon dioxide is the major component of the earth's crust); **oxidation**—exposure to oxygen produces oxides (e.g., rust); and, **carbonation**—the addition of carbon dioxide from the atmosphere to living things with hard outer shells. Carbon dioxide is released back into the atmosphere when the "limestone" remains of these creatures are exposed to ocean and atmospheric acid.

Go!

Give students ample time to perform the experiment described in Figure B on Journal Sheet #3. The pulverized sample usually yields more CO_2 because the acid comes in contact with more surface area of material.

Materials

plastic or glass petri dishes, paper towels, calcium carbonate (e.g., common blackboard chalk), hammer, mild acid (e.g., 1 molar HCl, available from laboratory supply house, or strong vinegar), medicine droppers, goggles

EA3 JOURNAL SHEET #3

CHANGING LAND

FIGURE B

<u>Directions</u>: (1) Wrap in a paper towel one half of the calcium carbonate (e.g., chalk) chips given to you by your instructor. (2) Use a hammer to gently pulverize the chips to powder. (3) Pour the powder into one of the petri dishes. (4) Place the remainder of the chips into the other petri dish. (5) One drop at a time, add a full medicine dropper of the acid given to you by your instructor to both samples of calcium carbonate. (6) Record your observations. Did the amount of carbon dioxide gas produced in the pulverized sample differ from that produced by the other sample?

CHANGING LAND

Work Date: ____/____/____

LESSON OBJECTIVE

Students will discuss how glaciers have changed the face of continents.

Classroom Activities

On Your Mark!

Show students a map of the United States and Canada and inform them that much of the North American continent was covered by a glacier that began to melt thousands of years ago. Tell them that the Great Lakes were carved out of the landscape by that giant moving mass of ice. Have students refer to pictures of glaciers available in most science textbooks or encyclopedias. Explain that a **glacier** is a large mass of ice formed by the compacting of annual snowfalls year after year during an Ice Age. Mention that scientists are not sure why Ice Ages occur. Some think they result from a slight shift in the earth's orbit. Others think they are caused by small changes in the sun's luminosity. No one knows for sure. Point out that the glacier in Antarctica covers nearly the entire land-mass (≈ 12,500,000 sq/km). Today, the Antarctic and other small glaciers are a natural reservoir for about 75 percent of our planet's fresh water.

Get Set!

Scientists have known for about one hundred years that glaciers move. But how they move has been a matter of some debate. Draw Illustration B to explain the currently accepted mechanism for glacial movement. Have students copy your drawing on Journal Sheet #4. Explain that as a glacier melts, it leaves behind all of the debris that became trapped in it when it froze. This debris is called **till**. Till contains debris from small pebble-sized particles to large boulders.

ILLUSTRATION B

The top of the glacier is called the **fracture zone** with **crevasses** as deep as 150 meters. Under pressure, the bottom of the glacier called the **flow zone** liquifies and flows downhill carrying the ice sheet with it. A glacier that moves downhill

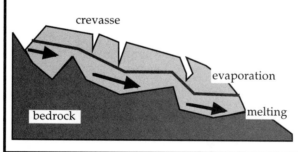

more rapidly than its lower end or "foot" melts is said to be "advancing." A glacier with a foot that melts back faster than the glacier moves downhill is said to be "receding." If the melting occurs at the same rate as the downward movement of the ice, then the glacier appears to be stationary.

Go!

Give students ample time to perform the experiment described in Figure C on Journal Sheet #4.

Materials

clay, small rocks and pebbles/nuts and bolts, ziplock baggies, water, tape

EA3 JOURNAL SHEET #4

CHANGING LAND

FIGURE C

Directions: (1) Pour about 100–150 ml of water into a ziplock sandwich baggie. (2) Drop several small rocks and pebbles, nuts and bolts, or other small objects into the baggie and zip it closed. DO NOT OVERFILL THE BAG! Use tape to be sure the bag is sealed against leakage. (3) Flatten some modelling clay into a sheet about 5 mm thick. (4) Place the baggie flat on the clay. (5) Pretend your hand is the heavy frozen top of a glacier. Press your palm down on the baggie and with light pressure "roll" the baggie across the clay. (6) Draw a picture of marks left on the clay. How is the pebble and waterfilled baggie like the flowing underside of a glacier? What would have happened to the pebbles and other objects if they had not been "locked" inside the baggie?

EA3 REVIEW QUIZ

Directions: Keep your eyes on your own work.
Read all directions and questions carefully.
THINK BEFORE YOU ANSWER!
Watch your spelling, be neat, and do the best you can.

CLASSWORK	(~40): _____
HOMEWORK	(~20): _____
CURRENT EVENT	(~10): _____
TEST	(~30): _____
TOTAL	(~100): _____

(A ≥ 90, B ≥ 80, C ≥ 70, D ≥ 60, F < 60)

LETTER GRADE: _____

TEACHER'S COMMENTS: _____

_____

CHANGING LAND

TRUE–FALSE FILL-IN: If the statement is true, write the word TRUE. If the statement is false, change the underlined word to make the statement true. *20 points*

_____ 1. <u>Physiologists</u> identify two major processes that contribute to the wearing down of the land: erosion and weathering.

_____ 2. <u>Erosion</u> is the process by which beaten rock and soil particles are moved from one place and deposited in another.

_____ 3. Loose rocks and soil are called <u>sediment</u>.

_____ 4. Sediment is usually deposited in layers that geologists call <u>strata</u>.

_____ 5. <u>The atmosphere</u> pulls on matter at the surface of the earth causing the downhill movement of sediment.

_____ 6. Wind <u>cannot</u> wear down solid rock.

_____ 7. A <u>glacier</u> is a large mass of moving ice and snow.

_____ 8. The Great Lakes on the border of the United States and Canada were carved out by a <u>great flood</u>.

_____ 9. <u>Chemical</u> weathering causes rocks to break into very small pieces without changing the chemical composition of the rocks themselves.

_____ 10. <u>Mechanical</u> weathering causes the minerals in rocks to change into new chemical substances.

List the factors that influence the time it takes for land features to become weathered. *3 points*

11. _____

12. _____

13. _____

PARAGRAPH FILL-IN: Use the vocabulary words you studied in this unit to best complete each sentence in the paragraph. *7 points*

    The surface of the earth is constantly being worn. The wearing of earth's surface can occur by (14) _____ or (15) _____. Erosion is a process that moves (16) _____ from one place to another. During (17) _____, however, materials change physically or chemically but remain in the same place. Landslides are a form of erosion due to (18) _____. Sand dunes are pushed across the surface of the desert by the (19) _____. At many times in earth's history (20) _____ have formed during Ice Ages. These massive ice sheets eventually melted and moved, carving out lakes and valleys across the surface of continents.

ROCKS AND MINERALS

TEACHER'S CLASSWORK AGENDA AND CONTENT NOTES

Classwork Agenda for the Week

1. Students will explain the rock cycle.
2. Students will examine the basic characteristics of igneous rocks.
3. Students will examine the basic characteristics of sedimentary rocks.
4. Students will examine the basic characteristics of metamorphic rocks.

Content Notes for Lecture and Discussion

The scientific revolutions of the 16th and 17th centuries, coincident with the discovery of the New World, marked the beginning of the modern science of geology. Early explorers documented new and different lifeforms and landscapes whose existence challenged the Biblical notion that the world remained static in the absence of divine intervention. Spurred on by the new "mechanical philosophies" of French mathematician **René Descartes** (b. 1596; d. 1650) and English scientist **Robert Hooke** (b. 1635; d. 1703), geologists began to accept the notion of change as basic to their understanding of the earth's crustal features. Hooke was among the first to vehemently argue—against the prevailing Biblical view—that fossils were the remains of living organisms that existed long ago. The work of the Danish anatomist and naturalist **Nikolaus Steno** (b. 1638; d. 1686) integrated fossil evidence with the observation that the crust of the earth was layered in strata. This knowledge allowed scientists to assign relative ages to the layers of the earth's crust and to establish the idea that change was continual. The works of Scottish geologists **James Hutton** (b. 1726; d. 1797) and **Charles Lyell** (b. 1797; d. 1875), culminating in the contributions of American physical geographer **William Morris Davis** (b. 1850; d. 1934), laid the foundation for the most concise description of rock formation and crustal change called the **rock cycle**.

Geologists base their notion of the rock cycle on the assumption that the processes shaping the earth's surface today are the same as those that were at work on earth in the distant and recent past. All rocks are thought to be derived from the cooling and crystallization of molten iron oxides (e.g., ferric oxide) and silicates (e.g., silicon dioxide) present at the active volcanic zones surfacing the planet. **Igneous rocks** formed in these regions provide the basic ingredients of all other rocks: namely, **minerals**. **Sedimentary rocks** are formed by the fragmenting, compaction, and cementing of these products. **Metamorphic rocks** are the transformed products of other rocks upon exposure to heat and pressure. Weathering, erosion, compaction, cementation, heat and pressure transform igneous rocks into the vast variety of mineralized combinations that sprinkle the planet. Excessive heat which exceeds the melting points of the minerals in rock forms new magma—the fluid of rock origination. The cycle begins again.

Robert Hooke, best known for popularizing the term "cell" to describe the basic unit of biology, published the first thorough description of mineral crystals in his book *Micrographia* in 1665. Hooke suggested that the underlying arrangement of the atoms and molecules in a crystal could be determined by examining the orientation of axes and faces of the crystal. For his work, Hooke is considered the "father of crystallography." The French minerologist and founder of modern crystallography, **René-Just Haüy** (b. 1743; d. 1822), developed the first system of classification used to categorize the varied forms of crystals. In 1784, he proposed that the smooth faces of a calcite crystal could be the result of stacking the individually cleaved layers of the crystal. The German minerologist **Friedrich Moh** (b. 1773; d. 1839) created the first hardness scale used to assist in the identification of crystallized minerals. Minerals can be differentiated according to their hardness, density, luster, streak, ductility, angle of cleavage, and chemical properties.

EA4 Content Notes (cont'd)

In Lesson #1, students will diagram the processes that give rise to the three basic types of rocks: igneous, sedimentary, and metamorphic.

In Lesson #2, students will examine the basic characteristics of igneous rocks and examine the properties of a common crystal commonly called salt.

In Lesson #3, students will examine the characteristics of sedimentary rocks and test for the presence of calcium carbonate, a compound prevalent in the hardened shells of mollusks.

In Lesson #4, students will compare and contrast the characteristics of metamorphic rocks to those of igneous and sedimentary rocks.

ANSWERS TO THE HOMEWORK PROBLEMS

Answers may vary but students should express the views that . . .

1. rocks are mixtures of minerals.
2. change is continuous and all rocks are subject to the same transformation processes.
3. a hot piece of magma can cool quickly in cold water, trapping air bubbles that decrease the mass of the rock per unit volume (e.g., pumice floats).
4. organisms which become fossils live and die on the earth's surface in the outer layers of the crust where sedimentary rocks are usually found.
5. it takes a lot of heat and pressure to form a metamorphic rock.

ANSWERS TO THE END-OF-THE-WEEK REVIEW QUIZ

1. mineral
2. true
3. have
4. hardness
5. usually
6. rock cycle
7. three
8. true
9. silicon dioxide
10. can

Answers will vary to questions A, B, and C. Students should rely on the methods used in Lessons #1, #2, #3, and #4 to phrase their replies.

(A) igneous rock
(B) metamorphic rock
(C) sedimentary rock
(D) weathering and erosion
(E) compaction
(F) melting
(G) heat and pressure
(H) sediment

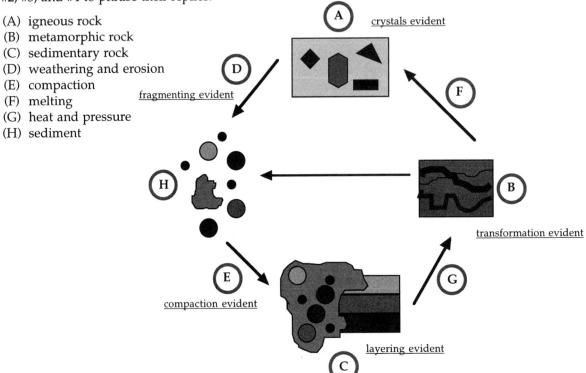

44

EA4 FACT SHEET

ROCKS AND MINERALS

CLASSWORK AGENDA FOR THE WEEK

(1) Explain the rock cycle.
(2) Examine the basic characteristics of igneous rocks.
(3) Examine the basic characteristics of sedimentary rocks.
(4) Examine the basic characteristics of metamorphic rocks.

Rocks are made of minerals. A **mineral** is a natural inorganic solid that has definite chemical composition. Figure A lists four common minerals found in the earth's crust. The last mineral on the list called quartz makes up about 70% of the earth's outer layer. All minerals are made of chemicals whose atoms have a definite arrangement. As a result, minerals form **crystals** that have identifiable shapes and qualities. Each of the minerals listed in Figure A has a definite crystalline structure and hardness that a trained geologist can identify. Figure B lists the relative hardness of 10 minerals that geologists use to identify minerals. The list is called **Moh's Hardness Scale** after the German mineralogist **Friedrich Moh** (b. 1773; d. 1839). A geologist who finds a mineral that cannot be readily identified scratches it against the minerals listed in Moh's Hardness Scale. If the unknown mineral is softer than orthoclase, for example, but harder than fluorite, then the mineral has a hardness of between 4 and 6. It could be apatite. The geologist will continue the identification process by measuring the density of the mineral. Since a mineral has definite chemical composition it has a specific density. The geologist who finds that the unknown mineral has a density of 3.2 g/cm^3 can be reasonably sure that it is apatite. Minerals are usually found mixed with other minerals. A mixture of minerals is called an ore. **Ores** contain both metals and nonmetals that can be removed in usable amounts. Copper, aluminum, and gold are three familiar substances that are purified from ores.

Geologists have been aware for some time that rocks can be changed from one form to another. The transformation of rocks is called the **rock cycle**. The processes that change rocks include (1) heating and melting, (2) cooling and crystallization, (3) erosion and weathering, and (4) compaction and cementation. The degree to which a particular rock is heated, cooled, fractured, and put under pressure determines how it is changed. There are three major families of rocks: igneous rocks, sedimentary rocks, and metamorphic rocks.

Igneous rocks are formed when hot magma cools and hardens on the surface of the earth or in the ocean. Some examples of igneous rocks are quartz, granite, and pumice. **Sedimentary rocks** are formed when sediments are deposited and compacted or cemented together. Some examples of sedimentary rock are sandstone, limestone, and all sorts of conglomerate or mixed rocks. Because sedimentary rocks are made near the earth's surface, they frequently contain **fossils**. **Fossils** are the hardened remains of

FIGURE A

Mineral	Chemical Name	Chemical Formula
calcite	calcium carbonate	$CaCO_3$
hematite	ferric oxide	Fe_2O_3
gypsum	hydrated calcium sulfate	$CaSO_4 \cdot 2H_2O$
quartz	silicon dioxide	SiO_2

FIGURE B

Number	Mineral	Can be scratched by ...
1	talc	any other solid
2	gypsum	fingernail
3	calcite	brass pin
4	fluorite	copper coin
5	apatite	window glass
6	orthoclase	quality steel
7	quartz	topaz
8	topaz	corundum
9	corundum	diamond
10	diamond	no other solid

organisms that were once alive. **Metamorphic rocks** are formed when chemical reactions due to tremendous heat or pressure change the chemical composition of the rock. Some examples of metamorphic rock are diamond, slate, and marble.

Homework Directions

Write two or more sentences that adequately answer each of the following questions.

1. How is a rock different from a mineral?
2. Why isn't the rock cycle a simple one-way process?
3. How might a rock that is lighter than water be formed?
4. Why are fossils usually found in sedimentary rocks?
5. Why are most metamorphic rocks found deep beneath the earth's surface?

Assignment due: _____

_____   _____   ____/____/____
Student's Signature   Parent's Signature   Date

EA4 Lesson #1

ROCKS AND MINERALS

Work Date: ____/____/____

LESSON OBJECTIVE

Students will explain the rock cycle.

Classroom Activities

On Your Mark!

Display **igneous** (e.g., quartz, granite, pumice), **sedimentary** (e.g., sandstone, limestone, any of the many cemented conglomerates), and **metamorphic rocks** (e.g., slate, marble, gneiss). Rock displays are available through laboratory supply houses should a wide variety be unavailable "in your own backyard." Give students 5 minutes to brainstorm the qualities of the rocks that they could use to categorize them (e.g., color, texture, etc.). What qualities would they use to group rocks together or distinguish between them? Since rocks are made of minerals, students may decide to group the rocks according to the minerals the rocks contain. Define a **mineral** as a natural inorganic solid with definite chemical composition. Refer them to Figure A on the Fact Sheet and explain the significance of each chemical formula for the minerals shown (e.g., minerals contain different kinds of atoms). Point out that a single rock can contain many different minerals depending upon how the rock was formed. For this reason, geologists have chosen to classify rocks according to how they are formed.

Get Set!

Review the processes of **volcanism**, **erosion**, and **weathering** studied in EA2 and EA3 of this volume. Draw Illustration A on the board and list off to the side the "processes" that cause rocks to change in form and composition. Help students to complete the diagram on Journal Sheet #1 by filling in the processes that result in the formation of each particular kind of rock.

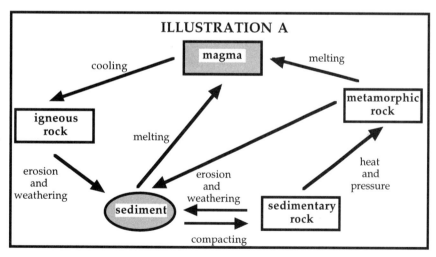

ILLUSTRATION A

Go!

Have students complete the activity and table described in Figure C on Journal Sheet #1.

Materials

samples of igneous (e.g., quartz, granite, pumice), sedimentary (e.g., sandstone, limestone, conglomerates), and metamorphic rocks (e.g., slate, marble, gneiss), magnifying glass

47

Name: ______________________________ **Period:**_______ **Date:** ___/___/___

EA4 Journal Sheet #1

ROCKS AND MINERALS

FIGURE C

Directions: (1) Examine with a magnifying glass each of the rock samples given to you by your instructor. (2) Refer to the illustration you completed above and classify each rock according to the process you think formed it. (3) Write a brief comment in the "comments" section explaining why you chose this particular process for each particular rock sample.

rock #	comment	process	rock class

ROCKS AND MINERALS

Work Date: ____/____/____

LESSON OBJECTIVE

Students will examine the basic characteristics of igneous rocks.

Classroom Activities

On Your Mark!

Display **igneous rocks** (e.g., quartz, granite, slate, marble, gneiss) and remind students that igneous rocks are formed by the cooling of molten magma. Explain that all matter (e.g., solid, liquid, or gas) is made up of **atoms** and that matter containing only one kind of atom is called an **element**. Refer to Illustration B and have students copy the list on Journal Sheet #2. The elements in this list are the most common elements in the earth's crust, elements that combine in the fiery furnace of volcanic vents to form a diverse variety of minerals. Remark that two elements—oxygen and silicon—make up about 75% of the earth's crust. This mineral is quartz: common sand. Refer students to Figure A on the Fact Sheet and the chart on Journal Sheet #2 to familiarize them with some of the most common minerals formed by fast or slow cooling of magma. The most characteristic feature of an igneous rock is its rich concentration of crystallized minerals. Fast cooling magma (e.g., plunged into water) forms small crystals. Slow cooling magma usually contains igneous rocks with larger crystals. Point out that atoms have very specific chemical properties which force them to combine in specific ways. A crystal, therefore, has a definite arrangement of atoms which gives it a very definite shape. Refer students to Figure D on Journal Sheet #2 and explain that there are six basic crystalline shapes. Crystals differ according to the different shapes and angles of the "faces." The face of a crystal is any two-dimensional side (e.g., a di has six identically shaped faces).

ILLUSTRATION B		
element	chemical symbol	percentage of earth's crust
oxygen	O	47
silicon	Si	28
aluminum	Al	8
iron	Fe	5
calcium	Ca	4
sodium	Na	3
potassium	K	3
magnesium	Mg	2

NOTE: These percentages are approximate. The earth's crust contains 92 elements. The others exist in much smaller amounts than those listed.

Get Set!

Before conducting the activity described in Figure E on Journal Sheet #2 students can sprinkle some common table salt onto a microscope slide and view it under a microscope. Table salt crystals composed of sodium chloride (chemical formula = NaCl) have a definite "cubical" shape.

Go!

Give students ample time to perform the activity described in Figure E on Journal Sheet #2. Review the procedure for finding the volume of an oddly shaped object (e.g., a rock) by water displacement. Show them how to fill a graduated cylinder with water, record the volume of the water and how much the water rises when the rock is submerged in the cylinder. They can weigh the rock on a balance and divide the mass of the rock by the volume of water it displaces to obtain the rock's density.

Materials

samples of igneous rocks, microscope and slides, table salt, graduated cylinders, beam balance, water

EA4 JOURNAL SHEET #2

ROCKS AND MINERALS

FIGURE D

tetragonal   hexagonal   triclinic

cubic   orthorhombic   monoclinic

COLOR AND DENSITY OF COMMON MINERALS

mineral	color	density
calcite	colorless	2.7
hematite	red, black	5.3
gypsum	white	2.3
quartz	colorless	2.6
galena	shiny gray	7.6
fluorite	violet	3.2
pyrite	gold	5.0

NOTE: These are the most common colors of these minerals. Some may have more than one color.

FIGURE E

Directions: (1) Examine each of the rock samples given to you by your instructor using a magnifying glass. (2) Measure the density of each rock by measuring its mass on a balance, then dividing that mass by the rock's volume. Measure the rock's volume by water displacement. (3) Refer to the chart titled COLOR AND DENSITY OF COMMON MINERALS and decide if your sample contains one or more of the minerals listed. (4) Write the name of that mineral in the correct column.

rock #	mass ÷ volume = density	Which minerals might be present in this rock?

ROCKS AND MINERALS

Work Date: _____/_____/_____

LESSON OBJECTIVE

Students will examine the basic characteristics of sedimentary rocks.

Classroom Activities

On Your Mark!

Prepare an extremely mild solution of 1 molar hydrochloric acid or purchase some from a laboratory supply house.

Display **sedimentary rocks** (e.g., sandstone, limestone, any of the many cemented conglomerates) and remind students that sedimentary rocks are formed by the erosion and weathering of other rocks followed by the compacting and cementing of the remaining bits and pieces. Sedimentary rocks have a variety of characteristic features. Some are simply clumps of sediment that have been cemented together. Others are pressed or compacted into layers giving the rock a "stratified" appearance. Still others contain the presence of fossilized remains (e.g., impressions or actual sea shells).

Get Set!

Display a group of sea or snail shells and explain that shells are made of calcium carbonate, the same compound that makes up the mineral calcite. Point out that sea shells belong to a biological group of animals called mollusks. Mollusks can absorb carbon dioxide from the air or water and combine it with calcium to produce calcium carbonate. Pouring acid on calcium carbonate splits the molecule and releases carbon dioxide bubbles back into the atmosphere. This method can be used to identify the presence of calcium carbonate in rocks that probably once contained the remains of living creatures. Blackboard chalk is also made of calcium carbonate.

Go!

Give students ample time to perform the activity described in Figure F on Journal Sheet #3.

Materials

samples of sedimentary rocks, 1 molar HCl, blackboard chalk, medicine droppers, magnifying glasses, goggles

EA4 JOURNAL SHEET #3

ROCKS AND MINERALS

FIGURE F

Directions: (1) Using a magnifying glass, examine each of the rock samples given to you by your instructor. (2) Write a brief comment about the most distinguishing feature of each sedimentary rock (e.g., layered, cemented, contains fossils, etc.). (3) Put a small drop of mild hydrochloric acid on the rock to test for the presence of calcium carbonate, a compound found in sea shells. Write "yes" if calcium carbonate is present and "no" if it is not. WEAR GOGGLES WHEN USING ACID!

rock #	comment	yes/no

ROCKS AND MINERALS

Work Date: _____/_____/_____

LESSON OBJECTIVE

Students will examine the basic characteristics of metamorphic rocks.

Classroom Activities

On Your Mark!

Prepare an extremely mild solution of 1 molar hydrochloric acid or purchase some from a laboratory supply house.

Display **metamorphic rocks** (e.g., slate, marble, gneiss) and remind students that metamorphic rocks are formed by the chemical changes that take place in other rocks when they are subjected to enormous amounts of heat and pressure. Metamorphic rocks usually have a smooth, shiny, or finely grained appearance that results from the partial melting and mixing of minerals in a rock. Glacial ice is considered a metamorphic rock because of the tremendous pressure exerted on the water crystals in the rock.

Get Set!

Have students compare, contrast, and discuss the physical appearance of metamorphic rocks and other rocks, igneous and sedimentary included.

Go!

Give students ample time to perform the activity described in Figure G on Journal Sheet #4. Summarize the unit by giving students ample time to finish Step #5 in the activity and open discussion of their conclusions. List the major similarities and differences between the three major classes of rocks on the board and have students copy the list on Journal Sheet #4.

Materials

samples of igneous, sedimentary, and metamorphic rocks, 1 molar HCl, medicine droppers, magnifying glasses, goggles, graduated cylinders, beam balance, water

EA4 JOURNAL SHEET #4

ROCKS AND MINERALS

FIGURE G

Directions: (1) Using a magnifying glass, examine each of the rock samples given to you by your instructor. (2) Write a brief comment about why each rock is a metamorphic rock. (3) Measure the density of each rock by measuring its mass on a balance, then dividing that mass by the rock's volume. Measure the rock's volume by water displacement. (4) Test for presence of calcium carbonate. (5) Write a paragraph to compare and contrast the properties of these rocks with those of the igneous and sedimentary rocks you have already examined.

rock #	comments	mass ÷ volume = density	yes/no

EA4 Review Quiz

Directions: Keep your eyes on your own work.
Read all directions and questions carefully.
THINK BEFORE YOU ANSWER!
Watch your spelling, be neat, and do the best you can.

CLASSWORK   (~40): _____
HOMEWORK   (~20): _____
CURRENT EVENT   (~10): _____
TEST   (~30): _____

TOTAL   (~100): _____
(A ≥ 90, B ≥ 80, C ≥ 70, D ≥ 60, F < 60)

LETTER GRADE: _____

TEACHER'S COMMENTS: _____
_____

ROCKS AND MINERALS

TRUE–FALL FILL-IN: If the statement is true, write the word TRUE. If the statement is false, change the underlined word to make the statement true. *20 points*

_____ 1. A <u>rock</u> is a natural inorganic solid that has a definite chemical composition.

_____ 2. The mineral <u>quartz</u> makes up about 70% of the earth's outer layer.

_____ 3. All minerals are made of atoms that <u>do not have</u> a definite arrangement.

_____ 4. The relative <u>size</u> of minerals is listed on Moh's Scale.

_____ 5. Minerals are <u>never</u> found mixed with other minerals.

_____ 6. The transformation of rocks is called the <u>mineral cycle</u>.

_____ 7. There are <u>five</u> major families of rocks.

_____ 8. <u>Fossils</u> are the hardened remains of organisms that once lived.

_____ 9. The chemical name of common sand is <u>calcium carbonate</u>.

_____ 10. Rock <u>cannot</u> be changed from one form to another.

ESSAY: List three things you could do to identify a rock. *3 points*

(A) _____
_____

(B) _____
_____

(C) _____
_____

Problem

Directions: Place the correct letter naming a process or rock at left that best describes each rock and arrow in the diagram. NOTE: Ignore the fact that this diagram is an incomplete picture of the rock cycle. *7 points*

(A) igneous rock
(B) metamorphic rock
(C) sedimentary rock
(D) weathering and erosion
(E) compaction
(F) melting
(G) heat and pressure
(H) sediment

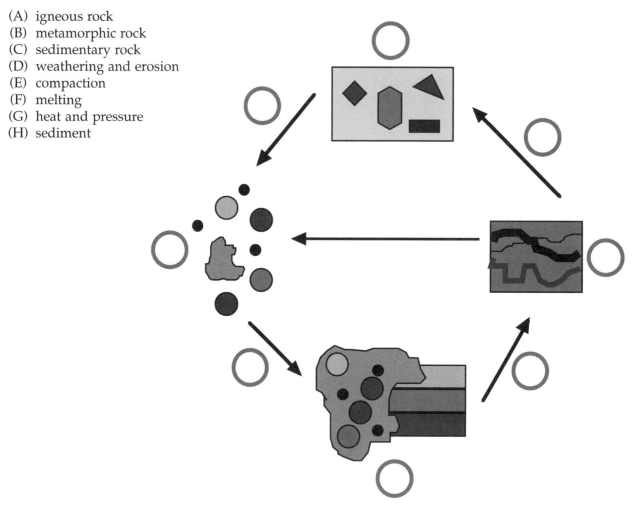

EARTH ORIGINS AND GEOLOGIC TIME

Teacher's Classwork Agenda and Content Notes

Classwork Agenda for the Week

1. Students will explain how gravity caused the formation of the earth and solar system.
2. Students will examine how to determine the age of rocks and rock strata.
3. Students will create a geologic time chart of the Cryptozoic Eon.
4. Students will create a geologic time chart of the Phanerozoic Eon.

Content Notes for Lecture and Discussion

The Danish anatomist and naturalist **Nikolaus Steno** (b. 1638; d. 1686) was the first to propose that fossils were the mineralized remains of organisms that inhabited the earth long ago. He noticed the similarities between strangely shaped rocks and the teeth and bones of animals alive today. Following his work and that of the English scientist **Robert Hooke** (b. 1635; d. 1703) who followed Steno's lead in defying the opposing view of the church in such matters, geologists concentrated their research in an effort to integrate fossil evidence into the history of rock strata. In Germany, minerologists employed by mining companies for the purpose of developing more efficient prospecting techniques laid the groundwork for an understanding of sequentially deposited rock formations. The German geologist **Abraham Gottlob Werner** (b. 1749; d. 1817) established the first physical explanation of the stratification process. He hypothesized that the earth was originally covered with water. Minerals suspended in solution became layered in sequential deposits as the water evaporated and receded. He linked strata to the history of the earth which supported Steno's notion of a "directional" development of the earth's crust. Werner's theory proved completely wrong and was replaced by the notions of **James Hutton** (b. 1726; d. 1797). In his book *Theory of the Earth* in 1795 Hutton introduced the **principle of uniformity** and the idea that rock types were formed by the differential heating of the earth by volcanic action and the complementary processes of weathering, erosion, and sedimentation. Although Werner's description of the mechanisms which contributed to the formation of earth's crust was replaced by that of Hutton, Werner's science of stratigraphy became firmly entrenched in the minds of geologists.

As geologists began to realize that minerals were not the key to an understanding of earth's history, fossils became the focus of attention. The English surveyor and mining prospector **William "Strata" Smith** (b. 1769; d. 1839) published *Strata Identified by Organized Fossils*, the first book to discuss the concept of the **index fossil**. The work of French comparative anatomists **Georges Cuvier** (b. 1769; d. 1832) and **Alexandre Brongniart** (b. 1770; d. 1837) produced a system for classifying fossils according to their similarities to existing lifeforms. The English geologist **Adam Sedgwick** (b. 1785; d. 1873) named the oldest fossil strata the Cambrian: now dated at 590 to 500 million years ago.

Since the beginning of the 20th century, the science of geology has—like other fields of science—diversified into a series of subdisciplines including stratigraphy, minerology, and petrology to name a few. Whereas the Irish physicist **William Thomson Kelvin** (b. 1824; d. 1907) first calculated the age of the earth by estimating the time it would take for earth's molten components to cool to their present temperature, the English geologist **Arthur Holmes** (b. 1890; d. 1965) pioneered **radioactive dating** methods. The latter have proved quite reliable in measuring the ages of ancient rocks and fossils.

EA5 Content Notes *(cont'd)*

In Lesson #1, students will explain how gravity caused the formation of the earth and solar system by determining the center of gravity situated within a random distribution of objects.

In Lesson #2, students will examine how index fossils and radioactive decay are used to determine the age of rocks and rock strata.

In Lesson #3 and Lesson #4, students will create a geologic time chart of the Cryptozoic Eon and the Phanerozoic Eon in order to visualize the enormous spans of geologic time that have existed throughout earth's changing history.

ANSWERS TO THE HOMEWORK PROBLEMS

Answers may vary but students should be able to see marked distinctions between the periods of their life.

ANSWERS TO THE END-OF-THE-WEEK REVIEW QUIZ

1. true	6. true	11. B	In their essay, students should mention Luis
2. Kepler	7. true	12. D	Alvarez, iridium, meteors or asteroids, dust
3. Newton	8. true	13. E	clouds blocking sunlight, disturbed food chain,
4. Descartes	9. true	14. A	and the crater in Yucatan, Mexico.
5. Whipple	10. index	15. C	

EA5 Fact Sheet

EARTH ORIGINS AND GEOLOGIC TIME

CLASSWORK AGENDA FOR THE WEEK

(1)   Explain how gravity caused the formation of the earth and solar system.
(2)   Examine how to determine the age of rocks and rock strata.
(3)   Create a geologic time chart of the Cryptozoic Eon.
(4)   Create a geologic time chart of the Phanerozoic Eon.

Human beings have wondered for thousands of years how the earth was formed. Ancient people used myths and legends to explain the beginnings of the earth, planets, and stars. But it wasn't until careful observations were made about the relationship between the earth, sun, and planets of our solar system that truly scientific ideas about earth's origin could be developed. The Polish astronomer **Nikolaus Copernicus** (b. 1473; d. 1543) was the first to show that the sun—and not the earth—was the center of the solar system. The German astronomer **Johannes Kepler** (b. 1571; d. 1630) described in detail the orbital motion of the planets around the sun. And the English mathematician-philosopher **Sir Isaac Newton** (b. 1642; d. 1727) explained how gravity held the planets in orbit around the sun. The French mathematician **René Descartes** (b. 1596; d. 1650) proposed in 1644 that the solar system was formed from a rotating disk of dust and gas. In 1755, the German philospher **Immanuel Kant** (b. 1724; d. 1804) supported this idea which later became known as the **nebular hypothesis**. The nebular hypothesis was revised recently by the American astronomer **Fred Lawrence Whipple** (b. 1906) and, today, is called the **dust-cloud hypothesis**. According to the dust-cloud hypothesis the solar system was formed in the following way: (1) The force of gravity and pressure created by light energy streaming into our solar system—perhaps from an exploding neighboring star—caused the disk of dust and small "planetesimals" to collapse. (2) Gravity was so intense at the center of the collapsing disk that **nuclear reactions** ignited a blazing sun. (3) The smaller, cooler spinning regions revolving around the sun were also drawn together by gravity to form each of the rotating planets. Modern **radioactive dating methods** suggest that the earth is more than 4.6 billion years old.

Geologists generally agree that forces like weathering and erosion that change the surface of our planet today were at work in the recent and distant past. This idea is called the **principle of uniformity**. They also agree that layers of rock and soil build up over long periods of time and that the oldest layers of rock are found at the deepest levels. This idea is called the **law of superposition**. Scientists use the principle of uniformity and the law of superposition to help estimate the age of **rock strata**. For example: If rocks and soil are deposited at the bottom of a lake at the rate of 1 centimeter every 1,000 years, then the lowest layer of sediment in a strata that is 10 centimeters thick at lake bottom is probably about 10,000 years old. Once scientists estimate the age of a particular rock strata they determine the age of the fossils in the strata. These fossils, called **index fossils**, are used to estimate the ages of other fossils found in deeper or more shallow layers of strata. **Fossils** are the mineralized remains of organisms that were once alive.

Geologists divide the history of the earth into two major **eons**: the Cryptozoic Eon and the Phanerozoic Eon. The **Cryptozoic Eon** spans the time from earth's formation to about 590 million years ago. The Cryptozoic Eon is divided into the **Hadean Era**, the **Archean Era**, and the **Proterozoic Era**. The Hadean Era includes the first 100 million years of earth's existence when the planet was still steaming hot and covered with poisonous gases bubbling from molten magma. The molecules necessary for the formation of life were probably manufactured during the Archean Era that lasted to about 2.5 billion years ago. During the Proterozoic Era small bacteria and other single-celled organisms began to form colonies and multicellular creatures. The appearance of numerous fossils at the start of the **Phanerozoic Eon** indicate that life really began to flourish at the start of this period. The Phanerozoic Eon is divided into the **Paleozoic Era**, the **Mesozoic Era**, and the **Cenozoic Era**. Each of these eras is characterized by the type of organisms that dominated that time (e.g., fish, reptiles, mammals).

Homework Directions

Directions: Draw a time chart like the one shown below that traces your life from your birth to the present and record one major event that occurred during each year. Divide the chart into major "eras" or time spans which can be categorized by a major event that took place during that time. For example "The Orange Street Era" might represent a 3-year period during which you and your family lived on Orange Street. "The Little League Years" might represent the time you played baseball in Little League.

13 years ago	I was born.	
12 years ago	I learned to walk.	The Orange Street Era
11 years ago	I got chicken pox.	

Assignment due: _____

_____   _____   ____/____/____
Student's Signature   Parent's Signature   Date

EARTH ORIGINS AND GEOLOGIC TIME

Work Date: ____/____/____

LESSON OBJECTIVE

Students will explain how gravity caused the formation of the earth and solar system.

Classroom Activities

On Your Mark!

Introduce Sir Isaac Newton's (b. 1642; d. 1727) **theory of universal gravitation**. Explain that according to Newton's theory—which was sufficiently useful in helping scientists to put men on the moon in the late 1960s—all massive objects attract one another. A "gravitational field" exists in space drawing all objects together at varying rates that depend on the individual masses of the objects and the distances between them. Use the information in the Teacher's Classwork Agenda and Content Notes to inform students that the ideas about the formation of the solar system have changed. Contrast the different theories.

Get Set!

Explain that all objects have a center of gravity. Perform the following demonstration: (1) Have a few students stand erect at the front of the class, arms to their sides. Tell them to remain "stiff as a board" until they feel themselves about to fall. (2) Place the ball of your palm at the center of their foreheads and gently push them backward. Make sure they step backward to prevent themselves from falling. (3) Ask them to describe what they were feeling at the moment they knew they were going to fall. Draw Illustration A on the board to show how their centers of gravity remained at the same point in their bodies while changing position with respect to the earth. Without their feet beneath them to act against the earth's pull they would surely fall.

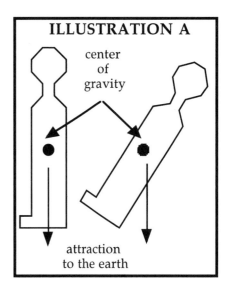

ILLUSTRATION A

center of gravity

attraction to the earth

Go!

Give students ample time to complete the activity described in Figure A on Journal Sheet #1.

Materials

ringstand and clamps, dissecting needles, string, nuts, cardboard, scissors

Name: _____   Period:_____   Date: ____/____/____

EA5 JOURNAL SHEET #1

EARTH ORIGINS AND GEOLOGIC TIME

FIGURE A

<u>Directions</u>: (1) Cut out a circle of cardboard at least 15 centimeters in diameter. (2) Randomly speckle the cardboard with dots and clumps to represent the dust, gas, and planetesimals from which the solar system might have formed. (3) Anchor a dissecting needle on a ringstand as shown. (4) Choose one of the larger planetesimals or dust clumps on the cardboard and hang the cardboard from the needle at that point. (5) Loop the string and suspended nut around the needle so that it hangs straight down. (6) Draw a line along the line of the hanging string. (7) Repeat Steps #4, #5, and #6 using the other dust-clumps and planetes-imals. (8) Note that all of the lines cross at the dust-cloud's "center of gravity." QUESTION: What object will be formed inside this region of space? Draw it.

EARTH ORIGINS AND GEOLOGIC TIME

Work Date: ____/____/____

LESSON OBJECTIVE

Students will examine how to determine the age of rocks and rock strata.

Classroom Activities

On Your Mark!

Begin with an explanation of the **principle of uniformity**, the **law of superposition**, and **index fossils** as described in paragraph #2 of the student Fact Sheet. Before the invention of **radiometric dating techniques**, paleontologists (e.g., scientists who study ancient life now extinct) estimated soil deposition rates and calculated the ages of strata based on the law of superposition. Fossils found in a particular layer are termed index fossils and are assumed to be the age of that layer. Fossils of the same kind found later could then be dated by comparing them to known index fossils.

Get Set!

Draw Illustration B on the board. Have students copy your diagram on Journal Sheet #2. Ask them to suppose that they have 16 kilograms of a mysterious "ELEMENT X." The substance is radioactive, giving off radioactive particles that you can detect using a **Geiger counter** invented by the German physicist **Hans Geiger** (b. 1882; d. 1945).

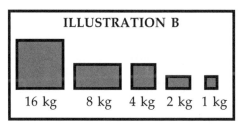

ILLUSTRATION B

16 kg 8 kg 4 kg 2 kg 1 kg

Atom after atom loses radioactive particles and the "click" of each particle is heard as it passes through the Geiger counter. By counting the clicks, we can measure the rate at which the lump of element X loses radioactive particles. Let's say we detect 1 click per second on our counter. Using this information we can calculate the number of atoms of element X that will disintegrate in a given amount of time. After doing a little bit of math, we determine that "half" of element X will be gone in one year at its present rate of decay. We say: "The **half-life** of element X is 1 year." Point out that neither intense heat nor cold will effect the rate of disintegration. So, in 4 years, there will be 1 gram of element X left in the rock sample.

Go!

Give students ample time to complete the activities described in Figure B and Figure C on Journal Sheet #2. Explain why the second activity is an analogy for how the age of rocks is determined by radioactive dating techniques. A person coming late to class simply needs to know the number of pennies you started with, the time now, and the time between shakes (e.g., the half-life) in order to be able to calculate what time it was when you first shook the box (e.g., the age of the fossil). The principle of uniformity allows geologists to make reasonable assumptions about the amount of radioactive material that was probably in a rock or dead organism when it was first deposited in fresh soil.

Materials

100 pennies or M&Ms™ per group, shoe boxes

EA5 JOURNAL SHEET #2

EARTH ORIGINS AND GEOLOGIC TIME

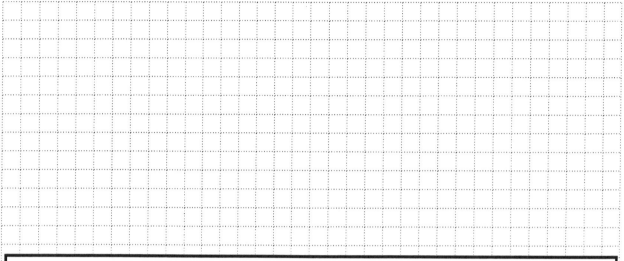

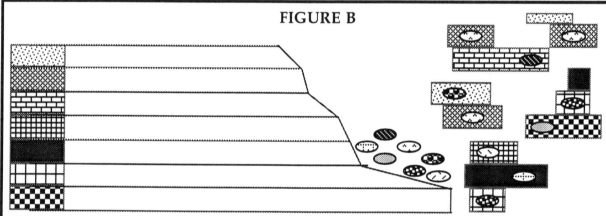

FIGURE B

Directions: Paleontologists have a record of the strata profile shown at left and the samples of several fossil-containing strata at right. They discover a jumble of fossils at the bottom of the riverbed as shown. Draw a duplicate copy of each fossil showing where it was originally deposited. If each strata was deposited over a period of 5 million years, how old is each fossil? NOTE: Assume that the top strata is still being deposited today.

FIGURE C

Directions: (1) Place 100 pennies or candies "heads up" in a shoe box or similar container. (2) Plot the number "100" on your graph to indicate that you started with 100 pennies (or candies). (3) Close and shake the box for ten seconds; then place it on the table and remove the pennies facing "tails up." (4) Plot the number of pennies left in the box on your graph. (5) Cover the box and shake it again four more times. Explain how someone who came late to class could tell you what time it was when you first shook the box.

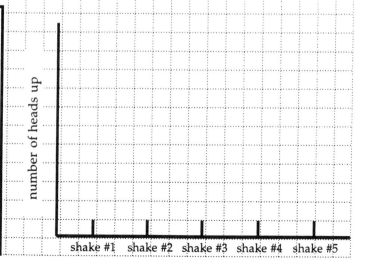

number of heads up

shake #1   shake #2   shake #3   shake #4   shake #5

EARTH ORIGINS AND GEOLOGIC TIME

Work Date: ____/____/____

LESSON OBJECTIVE

Students will create a geologic time chart of the Cryptozoic Eon.

Classroom Activities

On Your Mark!

Refer students to the Geologic Time Chart on Journal Sheet #3. Discuss the harsh conditions that probably existed on the primitive earth and which must have made life impossible during the first several hundred million years following the formation of the planet about 4.6 billion years ago. The atmosphere was most likely rich in poisonous **methane**, **carbon dioxide**, **ammonia**, and steaming **water vapor**. Point out, however, that these materials are the necessary precursors of the molecules that make up all living things: **carbohydrates**, **fats**, **proteins**, **genes** (e.g., DNA), and **liquid water**. Explain that **chlorophyll** may have been synthesized during the Archean Era allowing the evolution of plants to take place. Plants used carbon dioxide in the early atmosphere to manufacture carbohydrates and the oxygen that made the evolution of animal life possible. Multicellular life probably began during the Proterozoic Era.

Get Set!

Point out that the span of time being considered is enormous. Explain that there are 1 billion seconds in 32,000 years. It would take 147,200 years to count out 4.6 billion seconds.

Go!

Allow students ample time to complete the activity described in Figure D on Journal Sheet #3.

Materials

construction paper, scissors, tape, metric rulers

EA5 JOURNAL SHEET #3

EARTH ORIGINS AND GEOLOGIC TIME

GEOLOGIC TIME CHART OF THE CRYPTOZOIC EON		
Era	Years to Present	Dominant Life Form
Hadean	4.6 billion	none
Archean	3.8 billion	bacteria, algae
Proterozoic	2.5 billion	bacterial colonies

FIGURE D

Directions: (1) Cut strips of construction or butcher paper about 10 centimeters in width so that you can attach them end-to-end with glue or tape into a ribbon 5 meters long. (2) On your ribbon indicate that 1 millimeter (e.g., one thousandth of a meter) represents a span of 1 million years. (3) Draw lines along the width of paper segments to indicate the distance from the start of the ribbon to the times indicated in the GEOLOGIC TIME CHART OF THE CRYPTOZOIC EON (e.g., the start of the Archean Era is 800 million years or 800 millimeters from the start of the ribbon). (4) Write a paragraph on the ribbon and draw pictures to indicate what was happening on earth during the Hadean, Archean, and Proterozoic Eras. (5) Save the ribbon so that it can be used in Lesson #4.

EARTH ORIGINS AND GEOLOGIC TIME

Work Date: ____/____/____

LESSON OBJECTIVE

Students will create a geologic time chart of the Phanerozoic Eon.

Classroom Activities

On Your Mark!

Refer students to the Geologic Time Chart on Journal Sheet #4. Point out that the abundance of sea plants generating oxygen as a product of photosynthesis speeded the evolution of early sea creatures. The plants and animals that followed had to adapt to the earth's changing conditions resulting by variations in climate (e.g., ice ages) and landform (e.g., weathering and erosion). Remind them that continental drift contributed to the continual variation of species by slowly moving the continents into different climatic zones across the face of the planet.

Get Set!

Ask students to explain how they would account for the presence of tropical plant and jungle-dwelling dinosaur fossils beneath the glacial ice of Antarctica. They should conclude that this continent, having since moved as a result of continental drift, was once situated in an equatorial or tropical zone where these animals lived and died. Review the most widely accepted theory regarding the demise of the dinosaurs. The theory was first proposed by the American physicist **Luis Alvarez** (b. 1911; d. 1988). Alvarez and his research team discovered excessive amounts of the element iridium in samples of rock strata from the Late Cretaceous and Early Tertiary periods. Iridium is most commonly found in meteorites that come to earth from outer space. Alvarez suggested that the impact from a large meteor raised enormous volumes of dust into the atmosphere, preventing sunlight from hitting the earth in sufficient amounts. Disruption of the food chain following the mass extinction of millions of Mesozoic plant and animal species was the most probable cause of the dinosaurs' death. Recently, with the help of orbiting satellites, geologists have found a crater on the Yucatan Peninsula of Mexico which may be the impact site of that flaming asteroid.

Go!

Allow students ample time to complete the activity described in Figure E on Journal Sheet #4.

Materials

construction paper, scissors, tape, metric rulers

EA5 JOURNAL SHEET #4

EARTH ORIGINS AND GEOLOGIC TIME

GEOLOGIC TIME CHART OF THE PHANEROZOIC EON

Era	Period	Years to Present	Dominant Life Form
Paleozoic	Cambrian	590 million	seedless plants & cellular colonies
	Ordovician	500 million	shelled sea creatures
	Silurian	425 million	armored fish
	Devonian	405 million	amphibians & seed bearing trees
	Carboniferous	345 million	amphibians & land plants
	Permian	280 million	early reptiles
Mesozoic	Triassic	230 million	early dinosaurs & small mammals
	Jurassic	180 million	large dinosaurs
	Cretaceous	135 million	large dinosaurs
Cenozoic	Tertiary	63 million	small mammals
	Quarternary	1 million	large mammals & humans

FIGURE E

Directions: (1) Continue the activity started in Lesson #4 by drawing lines along the width of paper segments to indicate the distance from the start of the ribbon to the times indicated in the GEOLOGIC TIME CHART OF THE PHANEROZOIC EON. (2) Remember to write a paragraph on the ribbon and draw pictures to indicate what was happening on earth during the geologic eras and periods listed on the chart.

EA5 Review Quiz

Directions: Keep your eyes on your own work.
Read all directions and questions carefully.
THINK BEFORE YOU ANSWER!
Watch your spelling, be neat, and do the best you can.

CLASSWORK	(~40): _____
HOMEWORK	(~20): _____
CURRENT EVENT	(~10): _____
TEST	(~30): _____
TOTAL	(~100): _____

(A ≥ 90, B ≥ 80, C ≥ 70, D ≥ 60, F < 60)

LETTER GRADE: _____

TEACHER'S COMMENTS: _____

_____

_____

EARTH ORIGINS AND GEOLOGIC TIME

TRUE–FALSE FILL-IN: If the statement is true, write the word TRUE. If the statement is false, change the underlined word to make the statement true. *20 points*

_____ 1. The astronomer <u>Nicolaus Copernicus</u> was the first to show that the sun—and not the earth—was the center of the solar system.

_____ 2. The German astronomer <u>Sir Isaac Newton</u> described in detail the orbital motion of the planets around the sun.

_____ 3. The English mathematician-philosopher <u>Johannes Kepler</u> explained how gravity held the planets in orbit around the sun.

_____ 4. The French mathematician <u>Fred Lawrence Whipple</u> proposed in 1644 that the solar system was formed from a rotating disk of dust and gas.

_____ 5. The nebular hypothesis was revised recently by the American astronomer <u>René Descartes</u> and, today, is called the dust-cloud hypothesis.

_____ 6. The idea that forces like weathering and erosion that change the surface of our planet were at work in the recent and distant past is called the <u>principle of uniformity</u>.

_____ 7. The idea that layers of rock and soil build up over long periods of time and that the oldest layers of rock are found at the deepest levels is called the <u>law of superposition</u>.

_____ 8. Scientists use the principle of uniformity and the law of superposition to help estimate the age of <u>rock strata</u>.

_____ 9. <u>Fossils</u> are the mineralized remains of organisms that were once alive.

_____ 10. Fossils dated using the law of superposition are called <u>reference</u> fossils.

EA5 Review Quiz *(cont'd)*

MATCHING: Choose the letter of the group of organisms at right that was dominant during the geologic era at left. *5 points*

_____ 11. Hadean   (A) fish and amphibians

_____ 12. Archean   (B) no record of life forms

_____ 13. Proterozoic   (C) reptiles and small mammals

_____ 14. Paleozoic   (D) bacteria

_____ 15. Mesozoic   (E) colonies of single-celled organisms

ESSAY: Write a brief paragraph about the most widely accepted theory that explains the extinction of the dinosaurs. Include the name of the scientist who first proposed the theory and list the evidence that supports the theory. *5 points*

_____

_____

_____

_____

_____

_____

_____

_____

_____

_____

_____

_____

_____   _____   ___/___/___

Student's Signature   Parent's Signature   Date

THE OCEANS

TEACHER'S CLASSWORK AGENDA AND CONTENT NOTES

Classwork Agenda for the Week

1. Students will show how sonar is used to determine the depth of the oceans.

2. Students will map the patterns of the world's major ocean currents.

3. Students will explain the causes of waves and ocean tides.

4. Students will list and describe the components of sea water.

Content Notes for Lecture and Discussion

The ocean is a collection of large connected basins, smaller seas, bays, and gulfs that cover more than 71% of the planet's surface. The topography at the bottom of the ocean is much like that of the surface continents. Mountain ranges, volcanic vents, plains, plateaus, and deep cavernous trenches are scattered along and within the borders of the large crustal plates. The ocean teems with life and serves as the primary source of water for all living organisms on the planet. Water makes its way from sea to air to land and back to sea driven by the complementary processes of evaporation, condensation, and precipitation which comprise the **hydrological cycle**. Approximately 360,000 cubic kilometers of water evaporate from and return to the oceans every year compared to 70,000 cubic kilometers over the land. The ocean is the repository of minerals and organic remains washed from the continents by the continual mechanisms of erosion. Sediments found on the ocean floor are a mixture of **terrigenous** (e.g., land) **sediment**, **pelagic** (e.g., marine) **sediment**, **volcanic ash**, and **meteoritic matter**. Clay, silt, and sand are covered with **calcareous** (e.g., shelled organisms) and **siliceous ooze** from radiolarian (e.g., microscopic animal) and diatomaceous (e.g., microscopic plant) remains.

Sea voyages beginning in the 15th century and continuing to the present have contributed to the accumulation of vast amounts of oceanographic data. Perhaps the most complete oceanographic survey of the 19th century was conducted by the English oceanographer **John Murray** (b. 1841; d. 1914) aboard the *Challenger*. Murray summarized his findings in no less than 50 volumes of maps, charts, and descriptive text. Today, geocentric satellites use a variety of radar imaging techniques to map the oceans, their currents, and evolving physical and chemical properties.

Exploration of the ocean depths was impeded by the lack of appropriate technology until the 17th century. Following the invention of the submarine by the Dutch inventor **Cornelis Drebbel** (b. 1572; d. 1633) in 1620, the English scientist **Robert Hooke** (b. 1635; d. 1703) succeeded in devising a diving bell that allowed him to take samples of the relatively shallow sea floor. It was not until the middle of the 20th century, however, that the deepest ocean trenches could be explored. The Swiss scientist **Auguste Antoine Piccard** (b. 1884; d. 1962) devised a **bathyscaph** that allowed his son **Jacques Piccard** (b. 1922) to descend to the depths of the South Pacific ocean in 1953 and again in 1960. On the second descent Piccard submerged into the Marianas Trench near the island of Guam to a depth of 10,917 meters subjecting his craft to the enormous pressure of 15,000 pounds per square inch. One of the more exciting feats of deep ocean exploration was accomplished by the American oceanographer **Robert D. Ballard** (b. 1942) and his team. In 1985, the explorers sent a robot submersible named *Argo* into the dark waters of the cold North Atlantic to make the first video record of the luxurious ocean liner *Titanic* since it sank after striking an iceberg in 1912.

For the most part, the oceans remain a vastly unexplored realm. It was not until recently that scientists discovered life in the abyss thriving on geothermal energy along volcanic vents at

EA6 Content Notes *(cont'd)*

midocean ridges. Life in the depths was once thought impossible for lack of solar radiation to drive the photosynthetic-respiration cycle. This discovery has led scientists to the suggestion that life might exist on one or more of the moons orbitting the distant planet's Jupiter and Saturn. Is it possible that life exists beneath the ice of Europa or Titan, sustained by the geothermal energy of those giant satellites?

In Lesson #1, students will show how sonar is used to determine the depth of the oceans by graphic tabulated echo-sounding data.

In Lesson #2, students will map the patterns of the world's major ocean currents and examine how currents move.

In Lesson #3, students will explain the causes of waves and ocean tides.

In Lesson #4, students will list and describe the components of sea water and discuss impact of the modern pollutants on the ecological balance of the oceans.

ANSWERS TO THE HOMEWORK PROBLEMS

Students will discover that salt water does not freeze. Salt prevents the water molecules from crystallizing, thereby lowering the temperature at which the solution will turn solid. They should express this conclusion in the explanation of why the water at the North and South Poles remains liquid at sub-freezing temperature.

ANSWERS TO THE END-OF-THE-WEEK REVIEW QUIZ

1. true	6. continental shelf	11. F	16. D
2. Greece	7. abyss	12. B	17. A
3. true	8. are	13. G	
4. true	9. sonar	14. E	
5. true	10. 1.5	15. C	

Students should mention in their essay the important resources we get from the sea and how pollutants contaminate these resources.

In the diagram, students should draw both "neap tide" posts between the existing spring tide posts since the distance between low and high tide during neap tide is less than it is during spring tides.

EA6 Fact Sheet

THE OCEANS

CLASSWORK AGENDA FOR THE WEEK

(1) Show how sonar is used to determine the depth of the oceans.
(2) Map the patterns of the world's major ocean currents.
(3) Explain the causes of waves and ocean tides.
(4) List and describe the components of sea water.

The study of the world's oceans is called **oceanography**. Before the invention of the compass in early Greece around 500 BC the exploration of the ocean was limited to places near shore. Explorers avoided sailing out of sight of land fearing they would be unable to navigate themselves back home. In addition, early ships made of animal skins, reeds, and wood were not durable enough to survive the turbulence of deep sea and ocean waves. As navigation techniques improved and better ships were built explorers ventured out toward the undiscovered expanse of the open ocean wondering about the mysteries that lay beneath its surface. The first oar-driven submarine was constructed by the Dutch inventor **Cornelis Drebbel** (b. 1572; d. 1633) around 1620. Drebbel submerged his vessel to a depth of 15 feet beneath the surface of the River Thames in England. In 1801, the American engineer **Robert Fulton** (b. 1765; d. 1815) built the *Nautilus*. Fulton's hand-cranked, propeller-driven submersible was instantly recognized for its military value. The Irish-American inventor **John Holland** (b. 1840; d. 1914) experimented with the first modern submarine driven by an internal combustion engine in 1900. Since then, submarines powered by conventional fuels or nuclear energy have been used to explore the depths of the ocean.

As we move farther and farther from shore, the ocean bottom slopes downward for an average distance of 65 kilometers (e.g., about 40 miles) to an average depth of 61 meters (e.g., about 200 feet). It then plunges suddenly to an average depth of about 4 kilometers (e.g., about 2.5 miles). This slope is called a **continental shelf**. The bottom of the ocean is called the **abyss**. The deep ocean bottom is much like the land having plains, mountains, and valleys. It is covered with sediment containing slippery ooze, muds, and clays. In 1918, the French physicist **Paul Langevin** (b. 1872; d. 1946) invented a technique called **s**ound **na**vigation and **r**anging or **sonar** to help oceanographers map the abyss. Sound waves travelling at a speed of 1.5 kilometers per second through sea water are bounced off the ocean bottom. The distance to the bottom is determined by measuring the time it takes for the echo to return to the ship. If it takes 2 seconds for an echo to return, then the ocean depth at that point must be 1.5 kilometers.

The wind ripples the surface of the ocean with **waves** and drives the major **ocean currents**. Currents are also influenced by the earth's rotation which forces ocean water at the equator to flow toward the west. This flow produces clockwise movements of surface water in the Northern Hemisphere and counterclockwise movement in the Southern Hemisphere. Colder **countercurrents** have been discovered that flow across the bottom of the ocean against the movement of surface currents. Large tidal waves called **tsunamis** can be the result of severe storm or undersea earthquake activity. Tsunamis are responsible for the deaths of thousands of people every decade.

The gravitational pull of the sun and moon on the earth results in **tides**. High and low tides occur twice a day along every shoreline on our planet. When the moon is new or full and in direct alignment with the earth and sun, **spring tides** occur. Spring tides bring the waters closer inland and pulls them farther out to sea. When we have a quarter moon, **neap tides** occur. During neap tides the distance between low and high water lines is less than it is during spring tides.

The oceans supply us with more than 70 million metric tons of fish, crustaceans (e.g., shelled animals), and plants each year. Many of the food products we eat are made from **brown kelp**: a common seaweed. Gelatinous **algin** is extracted from kelp to make jellies and a variety of useful medicines. In addition, microscopic organisms called **plankton** generate much of the oxygen that all animals, including humans, need to breath. The ocean is rich in salt and other minerals. The elements iron, manganese, and nickel found in small lumps called **nodules** are being harvested every day from the sea.

Homework Directions

Perform the following experiment in order to determine why sea water at the earth's North and South Poles remains a liquid at subfreezing temperature. (1) Fill 2 small paper, plastic, or styrofoam cups halfway with tap water. *Do not use glass.* (2) Pour 1 tablespoon of table salt into one of the cups and stir. Label this cup with the word "salt." (3) Put both cups into the freezer and leave them there for 24 hours. (4) Remove the cups from the freezer and describe your observations. (5) Explain why sea water at the earth's North and South Poles remains a liquid at subfreezing temperature.

Assignment due: _____

_____   _____   ____/____/____
Student's Signature   Parent's Signature   Date

THE OCEANS

Work Date: _____/_____/_____

LESSON OBJECTIVE

Students will show how sonar is used to determine the depth of the oceans.

Classroom Activities

On Your Mark!

Use the information in the Fact Sheet and the Teacher's Classwork Agenda and Content Notes to give students a brief overview of the history of ocean exploration with an emphasis on the history of submarines. Review the discovery of the Italian physicist **Archimedes** (b. 287 B.C.; d. 212 B.C.) to explain how things float and perform the following demonstration shown in Illustration A: (1) Fill a 500 ml graduated cylinder with 250 ml of water. (2) Find the mass of a smaller 100 ml graduated cylinder using a balance. (3) Place the small cylinder into the larger one and record the amount of water displaced by the small cylinder. Note that the number of milliliters (ml) of water displaced is the same as the mass in grams of the small cylinder. Remind students that the definition of a gram is the amount of matter in one ml of pure water. (4) Remove the small cylinder and place a 20 or 50 gram brass weight into the cylinder. (5) Place the small cylinder with brass weight back into the larger cylinder. Note that 20 (or 50) additional grams of water are now displaced corresponding to the added mass of the brass weight used. Summarize **Archimedes' Principle**: A body placed in a fluid will displace a weight in fluid equal to its own weight. Point out that an object that can displace a volume of water with a larger mass than its own mass will float.

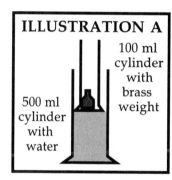

ILLUSTRATION A

500 ml cylinder with water

100 ml cylinder with brass weight

Get Set!

Show students how to construct an "eye-dropper submarine" like the one shown in Illustration B. Give them a few minutes to build their submarine and observe how they can control their submersible's ballast (e.g., the water in the dropper) by pressing on the sides of the plastic bottle. Since the air in the eye dropper is compressible, water can be forced in by squeezing the sides of the bottle. When pressure is released the compressed air in the dropper forces the water back out. Explain that submarines are responsible for much of the underwater oceanographic research that has been done during this century. Referring to the Fact Sheet, introduce the technique of sound navigation and ranging (e.g., **sonar**) first used by the French physicist **Paul Langevin** (b. 1872; d. 1946).

ILLUSTRATION B

Fill water to rim of bottle. Fill dropper half way before placing in bottle.

Go!

Give students ample time to complete the activity described in Figure A on Journal Sheet #1. Remind students to plot the points from "right to left" since West longitude readings increase from the Prime Meridian from East to West. ANSWERS: 10°W - 75 meters; 15°W - 150 m; 20°W - 750 m; 25°W - 1500 m; 30°W - 1500 m; 35°W - 300 m; 40°W - 150 m; 45°W - 75 m.

Materials

250 ml graduated cylinder, 500 ml graduated cylinder, 20 gram/50 gram brass weights, water, 2-liter plastic soda bottles, glass eye droppers.

EA6 JOURNAL SHEET #1

THE OCEANS

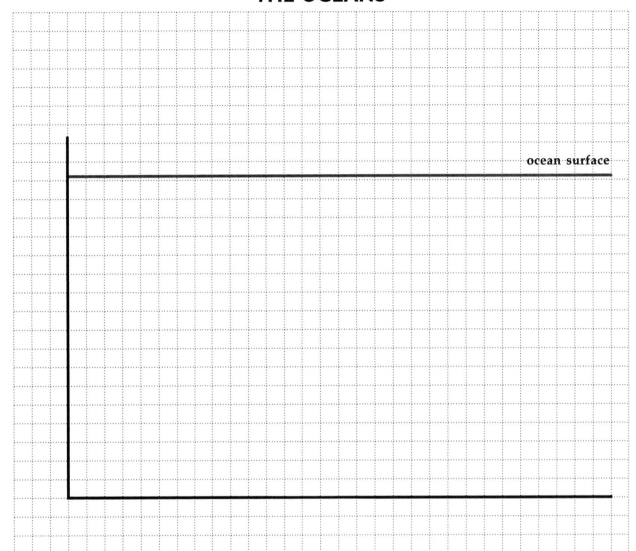

ocean surface

FIGURE A

Directions: (1) Examine the data presented in the SONAR TABLE. (2) Use the longitude readings to determine the expanse of the ocean to be plotted on the graph. (3) Calculate the depth of the ocean at each longitude by using the formula: (S x T) ÷ 2 = D; where S is 1.5 kilometers per second (e.g., the speed of sound in sea water), T is the time taken by the sound wave to echo back to the ship, and D is the distance to the bottom. (4) Plot the information on the graph to give a picture of the ocean bottom.

SONAR TABLE		
longitude	echo time	calculated depth
10° W	0.1 s	
15° W	0.2 s	
20° W	1.0 s	
25° W	2.0 s	
30° W	2.0 s	
35° W	0.4 s	
40° W	0.2 s	
45° W	0.1 s	

THE OCEANS

Work Date: ____/____/____

LESSON OBJECTIVE

Students will map the patterns of the world's major ocean currents.

Classroom Activities

On Your Mark!

Draw Illustration C on the board and have students copy the major features of the ocean floor. Note the trough created by the collision of the continental plate (e.g., at the shoreline) and the oceanic plate (e.g., in the abyss). Have students read paragraph #2 on their Fact Sheet. Then, using the information in the Teacher's Classwork Agenda and Content Notes they can lead a discussion of the depth characteristics of the continental shelf, trenches like the Marianas, and the abyss.

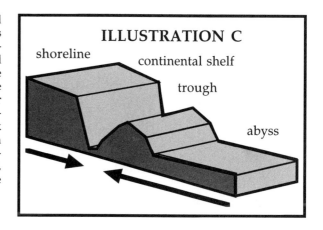

ILLUSTRATION C

shoreline · continental shelf · trough · abyss

Get Set!

Point out that ocean currents are driven by the wind which raises water into ripples along the ocean's surface. Explain that ocean currents are also influenced by the landmasses that obstruct the flow of water as well as the density and temperature of the water.

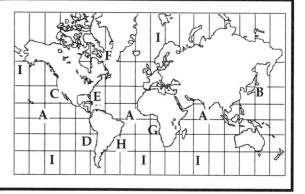

Major Ocean Currents

(A) Equatorial
(B) Japan
(C) California
(D) Humboldt
(E) Gulf Stream
(F) Labrador
(G) Benguela
(H) Brazil
(I) West Wind Drift

NOTE: The currents rotate clockwise in the North and counterclockwise in the South beginning at the equator.

Go!

Give students ample time to complete the activities described in Figure B on Journal Sheet #2. They will observe that the water is rippled by the artificial wind they create and that the currents on the surface circulate around the edges of the pan. Point out that the wind they created is analogous to the **Equatorial Currents** flowing west at the earth's equator. These currents are forced into a clockwise rotation in the Northern Hemisphere and a counterclockwise rotation in the Southern Hemisphere because of the earth's rotation and the landmasses that obstruct the waters' flow. Have students label the major ocean currents on the map provided on Journal Sheet #2.

Materials

tin pie plates, water, oregano or chili powder (or other cooking herb or powder that floats), paper or plastic straws, food coloring, 150 ml beakers, hot plate (if hot tap water is unavailable), table salt

EA6 Journal Sheet #2

THE OCEANS

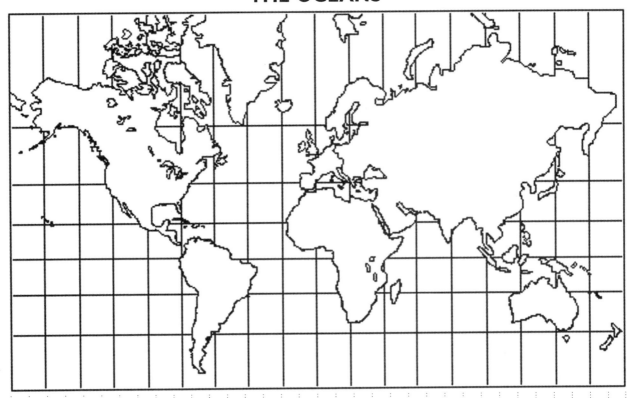

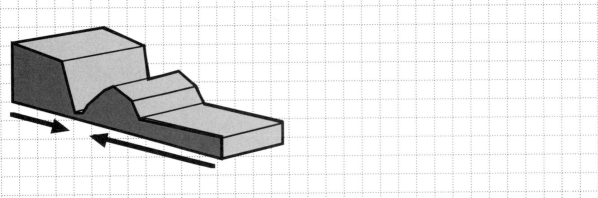

FIGURE B

Directions to Demonstration #1: (1) Fill a tin pie plate to the inner rim with cold tap water. (2) Sprinkle a teaspoon of herb (e.g., oregano, chili powder, etc.) over the surface. Using a plastic or paper straw, gently blow across the middle of the surface from one side of the pan. (3) Write a sentence to describe the patterns created by the motion of the herbs.

Directions to Demonstration #2: (1) If hot tap water is unavailable, use a hot plate to warm 50 milliliters of water. BE SURE TO EXERCISE PROPER SAFETY GUIDELINES IN THE USE OF THIS EQUIPMENT. WEAR GOGGLES. (2) Add a tablespoon of salt to the warm water and several drops of food coloring. Stir the mixture. (3) WEAR HEAT RESISTANT GLOVES and carefully pour several milliliters of the warm water into the same water-filled pie pan you used in Demonstration #1. (4) Write a sentence to summarize your observations about how this denser mixture behaves when poured into the cool water.

THE OCEANS

Work Date: _____/_____/_____

LESSON OBJECTIVE

Students will explain the causes of waves and ocean tides.

Classroom Activities

On Your Mark!

Begin with a presentation of Some Additional Facts About the Oceans. Hold up a globe so that students can view the Southern Hemisphere only. Turn the globe so that students can view only the Northern Hemisphere. Ask them to comment on the relative distribution of land and water in the two hemispheres. They should observe that the Southern Hemisphere is practically all water while the Northern Hemisphere has most of the land.

SOME ADDITIONAL FACTS ABOUT THE OCEANS		
ocean	area	average depth
Atlantic	87 million sq. km.	3,740 meters
Pacific	166 million sq. km.	4,190 meters
Indian	73 million sq. km.	3,870 meters
Arctic	12 million sq. km.	1,120 meters

Get Set!

Explain that ocean waves are raised by the winds and that the winds are produced by the differential heating of the earth's surface. This phenomenon will be discussed in Unit #7: *The Atmosphere*. Perform the following demonstration outside the classroom to illustrate the cause of ocean tides or, clear desks and chairs to make enough room. (1) Ask two students to hold hands and begin revolving around one another as in a dance. (2) Tell them to allow their free arms to go limp and loose. *Instruct the students to maintain self control so that they do not lose their balance and fall.* (3) Ask everyone to observe the direction in which the students' free arms move. Just like the skirt of a twirling ice skater their arms fly out and away from the students' clasped hands. The clasped hands are the center of rotation of this "two body" system. Now, explain that the earth and moon exert a mutual gravitational attraction on one another. The center of gravity between the earth and moon is located "off center" on the side of the earth facing the moon. As the two planetary bodies spin around one another centrifugal force causes the oceans to fly away from this center of gravity called a **barycenter**. Refer students to Figure C on Journal Sheet #3. Point out that these are views from above the earth's North Pole. As the earth rotates on its axis, the oceans and landforms move across the barycenter facing the moon. Landforms in line with the moon, earth, and barycenter on either side of the planet experience high tide because the oceans are being pulled away from the barycenter by centrifugal force.

Go!

Give students ample time to complete the activity described in Figure D on Journal Sheet #3. Explain that the **centrifugal force** pulling an object away from a gravitational source results from the momentum of the object as it tries to escape the pull of that source. The momentum of the water in the spinning cup keeps it pressed against the bottom of the cup as the cup spins. The bottom of the cup is, of course, being held in place by the string attached to the student's hand. The force being exerted through the string toward the hand is called **centripetal force**.

Materials

string, paper cups, water

Name: _____ **Period:** _____ **Date:** ___/___/___

EA6 JOURNAL SHEET #3

THE OCEANS

FIGURE C

During neap tide, the earth, moon, and sun are not aligned as the oceans fly away from the barycenter. During spring tide, the earth, moon, and sun are aligned, causing the "tidal bulge" to become exaggerated.

"X" marks the barycenter of the earth-moon rotation. Centrifugal force pulls the oceans away from this point. As the earth rotates on its axis, shorelines move in and out of the bulge resulting in two high and low tides each day.

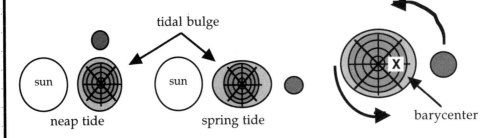

FIGURE D

Directions: This demonstration is best performed outside the classroom. (1) Punch 2 holes one-half inch below the rim of a sturdy paper cup (i.e., a paper coffee cup). (2) Tie the ends of a 2 meter length of string to the cup at the holes. (3) Fill the cup halfway with water. (4) Make sure that other students are standing clear. (5) Hold the string at the middle and begin to swing the cup in a circle above your head. Explain why the water does not spill from the cup.

THE OCEANS

Work Date: _____/_____/_____

LESSON OBJECTIVE

Students will list and describe the components of seawater.

Classroom Activities

On Your Mark!

Prepare for this lesson by cleaning and sterilizing a sufficient number of 250 ml beakers in dish soap and boiling water to insure that students will not be tasting contaminated water. Fresh table salt from the supermarket will also insure that proper health standards are maintained in the performance of the activity described in Figure E on Journal Sheet #4.

Instruct students to set up the experiment described in Figure E on Journal Sheet #4 (e.g., Steps # 1 through #4) and return to their seats.

Refer students to the table displaying the *Mineral Components of Seawater* on Journal Sheet #4. Point out that **sodium chloride** (e.g., common table salt) comprises over 85% of the minerals in seawater by weight. Review the processes of weathering and erosion that wash mineral-rich sediment out to sea via streams and rivers. Point out that the total volume of the oceans is about 1,370,000,000 cubic kilometers (e.g., 329,000,000 cubic miles). The total mass of salt in the oceans is about 50,000,000,000,000,000,000 kilograms (e.g., 50 million billion tons). If the oceans evaporated leaving the salt to be dispersed by the wind, the mineral would cover the planet with a layer of salt 45 meters (e.g., 150 feet) deep.

Get Set!

Give students several minutes to list the materials that are not in the table: namely, anything made of the organic material produced by living organisms. Have students discuss the conditions that would affect the "health" of the ocean. Discuss any recent newspaper or magazine articles reporting oil spills or the effects of other pollutants on the sea.

Go!

Give students ample time to complete the activity described in Figure E on Journal Sheet #4. They will discover that the water lining the inside of the baggie is not salty. Briefly explain the **hydrological cycle** (e.g., evaporation, condensation, precipitation) which will be demonstrated more thoroughly in Unit #9: *Introduction to Meterology*.

Materials

salt, water, desk lamps or sun, large ziplock baggies (clean and sterilized), 250 ml beakers

EA6 Journal Sheet #4

THE OCEANS

MINERAL COMPONENTS OF SEAWATER*		
mineral	grams per kilogram of seawater	% of total salt content
sodium ion	10.6	30.6
chlorine ion	19.0	55.0
sulfate ion	2.7	7.7
magnesium ion	1.3	3.7
calcium ion	0.4	1.2
potassium ion	0.4	1.1
other**	0.0	0.7
TOTAL	34.4	100.0

* all figures are rounded to the nearest tenth.
** the group "other" includes very small amounts of other minerals that make up less than one-tenth of a gram per kilogram of seawater

FIGURE E

Directions: (1) Pour a tablespoon of fresh table salt into the CLEAN AND STERILE 250 ml beaker provided by your instructor. (2) Add 100 ml tap water to the beaker and stir. (3) Place the beaker in a dry ziplock baggie and zip it closed to prevent leaks. (4) Place the beaker in warm direct sunlight or under a bright desk lamp. (5) Allow at least 30 minutes to pass and observe the inner lining of the baggie. If water droplets appear stuck to the lining, then you can open the bag and taste the droplets by wiping your finger across the bag's inner surface. Can you explain why the water droplets are not salty?

EA6 Review Quiz

Directions: Keep your eyes on your own work.
Read all directions and questions carefully.
THINK BEFORE YOU ANSWER!
Watch your spelling, be neat, and do the best you can.

CLASSWORK	(~40):	_____
HOMEWORK	(~20):	_____
CURRENT EVENT	(~10):	_____
TEST	(~30):	_____
TOTAL	(~100):	_____

(A ≥ 90, B ≥ 80, C ≥ 70, D ≥ 60, F < 60)

LETTER GRADE: _____

TEACHER'S COMMENTS: _____
_____

THE OCEANS

TRUE–FALSE FILL-IN: If the statement is true, write the word TRUE. If the statement is false, change the underlined word to make the statement true. *10 points*

_____ 1. The study of the world's oceans is called <u>oceanography</u>.

_____ 2. The compass was invented in early <u>Egypt</u> around 500 BC.

_____ 3. The first oar-driven submarine was constructed by the Dutch inventor <u>Cornelis Drebbel</u> around 1620.

_____ 4. In 1801, the American engineer <u>Robert Fulton</u> built the *Nautilus*: a hand-cranked, propeller-driven submersible.

_____ 5. The Irish-American inventor <u>John Holland</u> worked with the first modern submarine driven by an internal combustion engine.

_____ 6. The <u>abyss</u> slopes downward from the shore to a depth of 61 meters before plunging to a depth of about 4 kilometers.

_____ 7. The bottom of the ocean is called the <u>continental shelf</u>.

_____ 8. The contours of the deep ocean bottom <u>are not</u> similar to the contours of the land.

_____ 9. In 1918, the French physicist Paul Langevin invented <u>radar</u>.

_____ 10. Sound waves travel through sea water at an average speed of <u>10.5</u> kilometers per second.

ESSAY: Write a few sentences that explain why it is important to protect the oceans from industrial pollutants. *7 points*

_____

_____

_____

EA6 Review Quiz *(cont'd)*

MATCHING: Choose the letter of the ocean(s) at right that best matches the location of the major ocean surface current at left. *7 points*

_____ 11. Japan   (A) Atlantic and Pacific

_____ 12. Gulf Stream   (B) Western North Atlantic

_____ 13. Humboldt   (C) Western South Atlantic

_____ 14. Labrador   (D) Eastern Atlantic

_____ 15. Brazil   (E) North Atlantic

_____ 16. Benguela   (F) Western Pacific

_____ 17. Equatorial   (G) Eastern Pacific

DIAGRAM: The posts shown mark the location of high and low tide on a beach at spring tide. Draw two more posts to show the approximate locations of high and low tide at neap tide. *6 points*

low spring tide   high spring tide

THE ATMOSPHERE

TEACHER'S CLASSWORK AGENDA AND CONTENT NOTES

Classwork Agenda for the Week

1. Students will demonstrate that the atmosphere exerts pressure.
2. Students will examine the composition of the atmosphere.
3. Students will identify the layers of the atmosphere.
4. Students will demonstrate the greenhouse effect.

Content Notes for Lecture and Discussion

The study of the atmosphere began in the 17th century with the study of pumps. Siphons designed to drain water from mines and wells experienced difficulty as the excavations went deeper and deeper leading engineers and natural philosophers to speculate that air affected the operation of the pumps. **Galileo Galilei** (b. 1564; d. 1642) suggested that the "force of a vacuum" could only hold up the water to a maximum height. And it was not until **Evangelista Torricelli** (b. 1608; d. 1647) performed siphoning experiments with liquids of varying density that the culprit was shown to be the pressure exerted by the air surrounding the pump. Upon further experiment, Torricelli found that atmospheric pressure was less at higher elevations than it was at sea level. He demonstrated this phenomenon by sealing a glass tube at one end, filling it with liquid mercury, and inverting it into an open cup filled with the same heavy liquid. The level of mercury in the tube changed with the weather but was always lower—on average—at higher elevations. Torricelli explained that the air pressure exerted on the outer surface of the liquid in the cup was less at higher elevations allowing the weight of the mercury to cause it to fall. This fact suggested that the density of the air decreased continually as one went farther away from the earth's surface until there was no atmosphere at all. The idea that the earth was enveloped in a bubble of air surrounded by a heavenly vacuum was not a popular one. The works of **Aristotle** (b. 384 B.C.; d. 322 B.C.) which served as the foundation of accepted science by the Roman Catholic Church flatly denied the existence of a vacuum.

The elucidation of the chemical composition of the atmosphere was made possible by the discoveries of early chemists who identified the gases comprising the air. The discovery of **nitrogen**—which comprises 78% of the air by volume (76% by weight)—is credited to the English chemist **Joseph Priestley** (b. 1733; d. 1804). Nitrogen is the most abundant uncombined element in the atmosphere, existing mostly as a diatomic molecule (e.g., N_2). Priestley and the Swedish chemist **Karl Wilhelm Scheele** (b. 1742; d. 1786) shared recognition for the discovery of **oxygen**. It was Scheele, however, who first suggested that the air was composed primarily of two gases: one that supported combustion and another that did not.

On November 21, 1783, two French brothers—inventors **Joseph Michel** (b. 1740; d. 1810) and **Jacque Etienne Montgolfier** (b. 1745; d. 1799) made the first human ascent in a hot air balloon. The following century saw the accumulation of knowledge about the atmosphere as explorers tempted fate, harsh weather and asphyxiation, rising to altitudes of more than 29,000 feet. Thermometers and barometers were taken aloft to measure the temperature and pressure characteristics of the atmosphere at higher altitudes. During this century, modern instruments have allowed scientists to discern the layered structure of the atmosphere. Weather resulting from the differential heating of the earth's surface and the formation of convection cells occurs in the lowest layer of the atmosphere, the **troposphere**, which rises to about 50 kilometers (km). Convection cells do not occur in the **stratosphere** where the triatomic molecule **ozone** (e.g., O_3) absorbs ultraviolet radiation from the sun. The stratosphere is cool at the bottom and warmer at the top at an elevation of 85 km. Atop the stratosphere is the **mesosphere** which thins and cools to subfreezing temperatures. Above the mesosphere begins the **thermosphere**. At lower levels in the thermosphere (e.g., 100 to 500 km) atoms are ionized to form charged particles by the energetic particles and rays constituting the solar wind. At the top levels of the thermosphere gas molecules can travel several hundred meters before striking another particle. At sea level, atmospheric gas molecules are on average one-hundred-millionths of a meter apart. Nevertheless, charged particles streaming into the atmosphere along the **Van Allen Radiation Belts**—discovered by **James Alfred Van Allen** (b. 1914) in 1958—excite atmospheric gases to produce the **auroras** at both of the poles.

The energy absorbed by the atmosphere originates in the sun. About 25% of the solar radiation striking the ocean of air surrounding the planet is reflected back into space by the air and clouds. About 5% is reflected back into space by the surface of the planet mostly at the poles. About 20% is absorbed by the air and clouds while the remaining 50% is absorbed at the surface. Energy absorbed at the surface is reradiated back into the atmosphere in the long-wave or infrared range. Scientists fear that an accumulation of dust and some gases (e.g., carbon dioxide) will hold infrared radiation near the surface thereby warming the troposphere. This phenomenon is called the **greenhouse effect**. In addition, aerosol pollutants like chlorofluorocarbons (e.g., CFC) upset the synthesis and decomposition cycle of ozone. Recently, satellite imaging techniques have detected an expanding region of ozone depletion at the South Pole. This depletion could allow increased levels of ultraviolet radiation to warm the surface of the planet, melt significant amounts of glacial ice and raise sea level to catastrophic depths.

In Lesson #1, students will demonstrate that the atmosphere takes up space and exerts pressure.

In Lesson #2, students will examine the composition of the atmosphere.

In Lesson #3, students will identify the layers of the atmosphere by graphing pressure, temperature, and elevation.

In Lesson #4, students will demonstrate the greenhouse effect.

ANSWERS TO THE HOMEWORK PROBLEMS

Student should observe a rise in the level of water in the straw over several days as atmospheric oxygen combines with iron in the nail to form rust. Some students may even comment that the water rises only one-fifth the level of the straw and no more, demonstrating that oxygen comprises about 20% of the air around us.

ANSWERS TO THE END-OF-THE-WEEK REVIEW QUIZ

1. atmosphere	6. true	11. C
2. is	7. true	12. E
3. 14.7	8. true	13. A
4. bar	9. true	14. B
5. barometer	10. true	15. E

EA7 FACT SHEET

THE ATMOSPHERE

CLASSWORK AGENDA FOR THE WEEK

(1)   Demonstrate that the atmosphere exerts pressure.
(2)   Examine the composition of the atmosphere.
(3)   Identify the layers of the atmosphere.
(4)   Demonstrate the greenhouse effect

The earth is surrounded by a thin blanket of air we call the **atmosphere**. Air is like all other forms of matter and is made of atoms and molecules that have mass. Because air has mass it is attracted to earth by gravity. The atmosphere rises to about 500 kilometers (e.g., 310 miles) above the earth's surface exerting pressure on everything in it. Air pressure at sea level is about 1,000,000 dynes per square centimeter or about 14.7 pounds per square inch. This measure of atmospheric pressure is called one **bar**. The Italian scientist, **Evangelista Torricelli** (b. 1608; d. 1647) designed the first **barometer** in 1643 to measure the force being exerted by the atmosphere. Upon taking his barometer to higher elevations, Torricelli discovered that there was less atmospheric pressure surrounding him. He reasoned that the air became less dense as one went higher into the atmosphere. In fact, three-quarters of the earth's atmosphere lies between 8 and 16 kilometers (e.g., 5 to 10 miles) above sea level.

The lowest layer of the atmosphere is called the **troposphere**. The troposphere rises to about 16 kilometers above sea level. The average temperature of the troposphere varies from about –24°C (e.g., –75°F) to 24°C (e.g., 75°F). All weather—wind, rain, snow, and hail—occurs in the troposphere. The **stratosphere** is the next layer of the atmosphere. It rises to about 50 kilometers (e.g., 31 miles) above sea level and has an average temperature range between 10°C (e.g., –50°F) and 0°C (e.g., 32°F). The stratosphere protects life on the surface of the planet by absorbing dangerous **ultraviolet radiation** from the sun. Today, scientists fear that air pollution may be weakening the concentration of a stratospheric gas called **ozone** which absorbs this deadly radiation. Without the protection of ozone in the stratosphere, the atmospheric temperature could increase as it absorbs excess solar energy. The **mesosphere** is the third layer of the atmosphere. The mesosphere rises to 85 kilometers (e.g., 53 miles) and cools slowly as elevation increases from about 0°C to –90°C (e.g., –122°F). Above the mesosphere is the **thermosphere**. Atoms and molecules in the thermosphere absorb solar energy and become more energetic. As a result, the temperature in the thermosphere increases with increasing altitude. However, as matter becomes scarce in outer space, surfaces facing away from the sun get icy cold while surfaces facing toward the sun become extremely hot. Astronauts working in space must wear specially designed spacesuits to withstand sudden drastic changes in temperature as they move about.

The atmosphere is made primarily of **nitrogen gas**. Nitrogen makes up about 78% of the air around us. **Oxygen gas** makes up about 21%. The remaining 1% is a combination of other gases: **argon**, **carbon dioxide**, and more.

Beyond the earth's atmosphere is a vast magnetic field that attracts particles of matter streaming toward us from the sun. This field of magnetism was discovered by American physicist **James Alfred Van Allen** (b. 1914) in 1958. There are two **Van Allen Radiation Belts** surrounding the earth like donuts made of dust and gas wrapped around a golf ball. The first belt extends from 1,000 to 5,000 kilometers above the equator. The second is a cloud of charged particles extending from 15,000 to 25,000 kilometers. Some of the solar radiation pouring into our atmosphere rides the Van Allen Belts like a surfer on a wave toward the North and South Poles. As the radiation hits the top of the atmosphere colorful lights stream across the sky at the poles. These "light shows" are called the **Aurora Borealis** (e.g., north) and **Aurora Australis** (e.g., south).

Homework Directions

Perform the following experiment in order to demonstrate that the atmosphere is about one-fifth oxygen gas. (1) Use a marking pen to divide the length of a clear plastic straw into five equal sections. (2) Insert a wet iron nail into the straw and cover the head of the nail and straw with clay. Make sure that no air can leak into the straw. (3) Use tape to secure the straw to the inner side of a glass of water as shown in the diagram. Make sure that the straw is not touching the bottom of the glass. (4) Record the height of the water in the straw every day for the next two days. (5) Note that the nail is made of iron (element = Fe) and that rust is made of iron oxide (formula = Fe_2O_3). Write a paragraph that explains why this demonstration proves that the atmosphere contains oxygen gas.

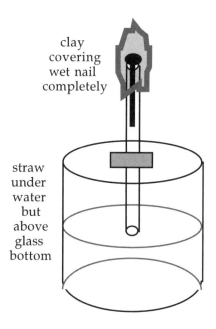

clay
covering
wet nail
completely

straw
under
water
but
above
glass
bottom

Assignment due: _____

_____   _____   ____/____/____
Student's Signature   Parent's Signature   Date

THE ATMOSPHERE

Work Date: ____/____/____

LESSON OBJECTIVE

Students will demonstrate that the atmosphere exerts pressure.

Classroom Activities

On Your Mark!

Begin with a demonstration that air takes up space as shown in Illustration A. (1) Prepare two flasks: one with a glass or plastic funnel in a 1-hole rubber stopper and another with a funnel and glass tube through a 2-hole rubber stopper. (2) Use a beaker to pour water into each flask. If poured quickly the water in the first funnel will not flow into the flask. Ask students to explain why the water does not flow into the flask. Answer: The air inside the flask takes up space and cannot escape. The tube in the second flask gives the air a way out. So water can flow into the second flask.

ILLUSTRATION A

Get Set!

Perform the demonstration shown in Illustration B to demonstrate that air exerts pressure: (1) Fill a test tube with water and place a playing/index card over the top. (2) Slowly invert the test tube over a sink or pan while holding the card in place. With practice, you should be able to keep the card from falling. The pressure of the atmosphere will hold it in place as illustrated. Ask: "What force is holding up the card against the force of gravity trying to pull the water down?" Answer: atmospheric pressure. Perform the demonstration shown in Illustration C: (1) Invert a water-filled test tube into a beaker/glass of water and allow a small amount of water to spill from the tube creating a space filled with air inside the tube. (2) Ask: "What will happen to the water in the tube if I start to lift the test tube off the bottom of the beaker?" Do it. Students will observe that the level of the water in the tube does not change. Describe the experiments of **Evangelista Torricelli** (b. 1608; d. 1647) mentioned in the Teacher's Classwork Agenda and Content Notes.

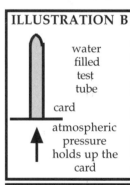

ILLUSTRATION B

water filled test tube

card

atmospheric pressure holds up the card

Go!

Have students perform the activity described in Figure A on Journal Sheet #1. Explain that heating the can caused the air inside the can to expand and escape. When the can was inverted into the bowl of cool water, the remaining air inside contracted. This allowed the atmospheric pressure outside the can to crush it.

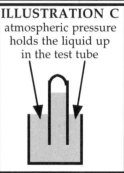

ILLUSTRATION C
atmospheric pressure holds the liquid up in the test tube

Materials

test tubes, beakers, index/playing cards, 1-holed stoppers, 2-holed stoppers, glass or plastic funnels, glass tubing, soda cans, Bunsen burners, ringstands and clamp, tongs, Ehrlenmeyer flasks, water

EA7 JOURNAL SHEET #1

THE ATMOSPHERE

FIGURE A

<u>Directions</u>: (1) Pour several milliliters of water into a soda can. (2) Place the soda can onto a ringstand and secure it loosely with another ring and clamp so that it cannot be toppled but can be removed easily with tongs. (3) Fill a small bowl with water. (4) Turn on the Bunsen burner and wait until steam starts to escape from the can. (5) Count slowly to 20. (6) Holding the can securely with tongs, <u>quickly</u> lift and invert the can upside down into the bowl of cool water.

GENERAL SAFETY PRECAUTIONS

Be sure you are familiar with the proper use of a Bunsen burner. Wear goggles. Do not touch any part of the equipment without heat-resistant gloves or tongs.

THE ATMOSPHERE

Work Date: ____/____/____

LESSON OBJECTIVE

Students will examine the composition of the atmosphere.

Classroom Activities

On Your Mark!

Refer to the Teacher's Classwork Agenda and Content Notes and begin with a brief discussion of the work of **Joseph Priestley** (b. 1733; d. 1804) and **Karl Wilhelm Scheele** (b. 1742; d. 1786) who discovered the gases **nitrogen** and **oxygen**. Refer students to Table A on Journal Sheet #2 and point out that the atmospheric gases listed in the column labelled "Today's Atmosphere" can be measured using instruments that are sensitive to different chemical compounds. Explain that their homework will verify the presence of one of the major components of the atmosphere: oxygen.

Get Set!

Point out that scientists suspect that the atmosphere has changed since our planet formed 4.5 billion years ago. Review the theories that explain earth's formation discussed in Lesson #1 of Unit 5: *Earth Origin and Geologic Time*. Explain that the gases listed in Table A on Journal Sheet #2 in the column designated "Ancient Earth" are the same gases that spew from volcanoes today. Since the ancient earth was probably covered with active volcanoes after its formation during the Cryptozoic Eon, it is likely that the gases listed in this column made up the earth's primitive atmosphere.

Go!

Give students sufficient time to brainstorm the events that must have occurred on earth to change the contents of the atmosphere. As the discussion proceeds, circulate and review with individual groups the role played by plants (e.g., through the process of **photosynthesis**) in the evolution of the primitive atmosphere. Write the chemical equation for photosynthesis on the board and have students copy it onto Journal Sheet #2: $6CO_2 + 6H_2O \rightarrow C_6H_{12}O_6 + 6O_2$. Explain that carbon dioxide was consumed by plants and used to manufacture oxygen. Point out that the evolution of life resulted in the reduction of other gases whose molecules were used in the manufacturing of the building blocks of life: carbohydrates from carbon dioxide and water, proteins from ammonia-containing nitrogen and carbohydrate-related molecules.

Give students time to complete the activity described in Figure B on Journal Sheet #2. Ask students to explain their observations. *Explanation:* The candle will consume all of the oxygen in the cylinder and go out. The amount of carbon dioxide produced takes up less volume than the oxygen consumed. Atmospheric pressure on the surface of the water in the beaker will force the water up the cylinder.

Materials

birthday candles, matches, 500 ml beakers or bowls, 100 ml graduated cylinders

EA7 JOURNAL SHEET #2

THE ATMOSPHERE

TABLE A

ATMOSPHERIC GAS	ANCIENT EARTH	TODAY'S ATMOSPHERE
carbon dioxide (CO_2)	92.2%	0.03%
nitrogen (N_2)	5.1%	78.1%
hydrogen sulfide (H_2S)	0.2%	0.0%
sulfur dioxide (SO_2)	2.3%	0.0%
methane (CH_2)	0.1%	0.0%
ammonia (NH_2)	0.1%	0.0%
oxygen (O_2)	0.0%	20.9%
argon (Ar)	0.0%	0.9%

FIGURE B

<u>Directions</u>: (1) Melt several drops of wax off the bottom of a birthday candle allowing the wax to drip into the bottom center of a 500 ml beaker or bowl. (2) Quickly place the candle into the melted wax securing the candle to the bottom of the beaker or bowl as the wax cools and hardens. (3) Use the 100 ml graduated cylinder to slowly pour 50 milliliters of water into the 500 ml beaker or bowl. (4) Light the birthday candle. (5) Invert the empty 100 ml cylinder and gently lower it down over the lit candle to the bottom of the beaker. (6) Record your observations.

GENERAL SAFETY PRECAUTIONS

DO NOT TOUCH THE HOT WAX WITH BARE FINGERS. Wear goggles.

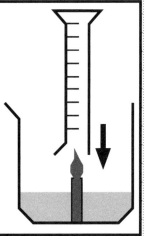

THE ATMOSPHERE

Work Date: ____/____/____

LESSON OBJECTIVE

Students will identify the layers of the atmosphere.

Classroom Activities

On Your Mark!

Use the information on the Teacher's Classwork Agenda and Content Notes and student Fact Sheet to help students list the important characteristics of each layer of the atmosphere: **troposphere**, **stratosphere**, **mesosphere**, and **thermosphere**.

Get Set!

Display photographs of earth taken from space showing cloud formations and the thin envelope of air enveloping the planet. Inform students that the radius of the earth is nearly 6,000 kilometers while the atmosphere rises to barely 100 kilometers. Display a globe and measure the radius of the globe. Ask students the following question: "If this globe were the actual size of the earth, how thick would the atmosphere be?" *Answer:* 100/6,000 or 0.01667 times the radius of the globe [e.g., if the radius of the globe is 20 centimeters, the atmosphere surrounding the globe would average 0.3334 centimeters (e.g., 3.3 millimeters)].

Go!

Assist students in graphing the information in Table B on Journal Sheet #3.

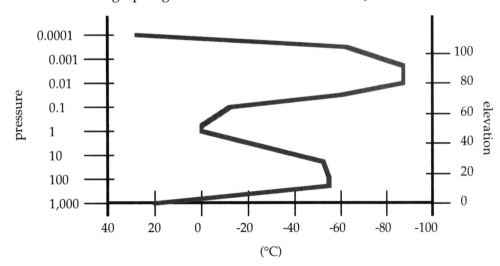

Materials

globe of earth, rulers, magazine photos of earth taken from space

EA7 JOURNAL SHEET #3

THE ATMOSPHERE

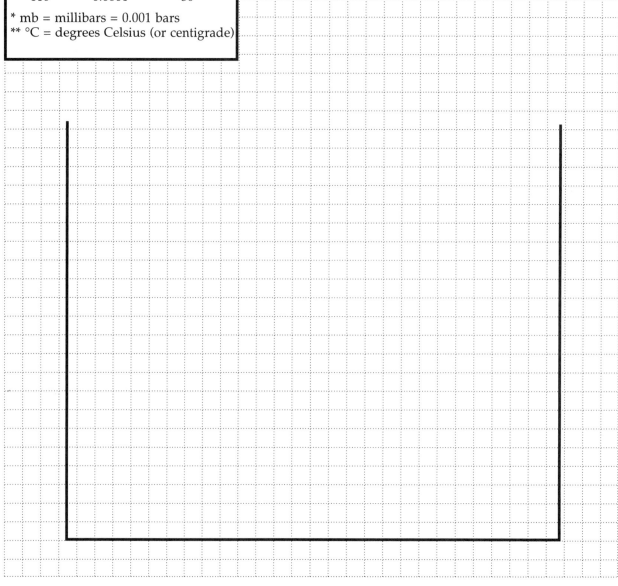

TABLE B

elevation	pressure	temperature
(km)	(mb)*	(°C)**
0	1,000	20
10	500	-55
20	100	-55
30	10	-50
40	1	0
50	1	0
60	0.1	-10
70	0.05	-60
80	0.01	-90
90	0.001	-90
100	0.0005	-60
110	0.0001	30

* mb = millibars = 0.001 bars
** °C = degrees Celsius (or centigrade)

THE ATMOSPHERE

Work Date: ____/____/____

LESSON OBJECTIVE

Students will demonstrate the greenhouse effect.

Classroom Activities

On Your Mark!

Prepare the thermometer-rubber stopper assemblies shown in Figure C on Journal Sheet #4. Use glycerine to ease the insertion of the thermometers.

Use the information on the Teacher's Classwork Agenda and Content Notes to give a brief lecture about the **greenhouse effect**. If available, show students satellite-imaging photographs taken from national magazines or scientific journals of the ozone hole developing over the Antarctic. Display pictures of smog-ladened cities, and industrial pollutants being emitted by the smokestacks of manufacturing facilities.

Get Set!

Draw Illustration D on the board and have students copy it onto Journal Sheet #4. Explain how the greenhouse effect occurs. Point out that solar radiation enters a glass greenhouse as shortwave energy. The energy is absorbed by the atoms and molecules of the objects inside the greenhouse. The atoms and molecules inside the greenhouse give off long-wave infrared radiation that cannot escape through the glass. The temperature in the greenhouse rises.

ILLUSTRATION D

Go!

Assist students in performing the activity described in Figure C on Journal Sheet #4. They will discover that the temperature inside the flask with the burnt matches rises more quickly.

Materials

matches, 250 ml Ehrlenmeyer flasks, 1-hole rubber stoppers, thermometers, desk lamps or bright sun

Name: _____ **Period:** _____ **Date:** ____/____/____

EA7 JOURNAL SHEET #4

THE ATMOSPHERE

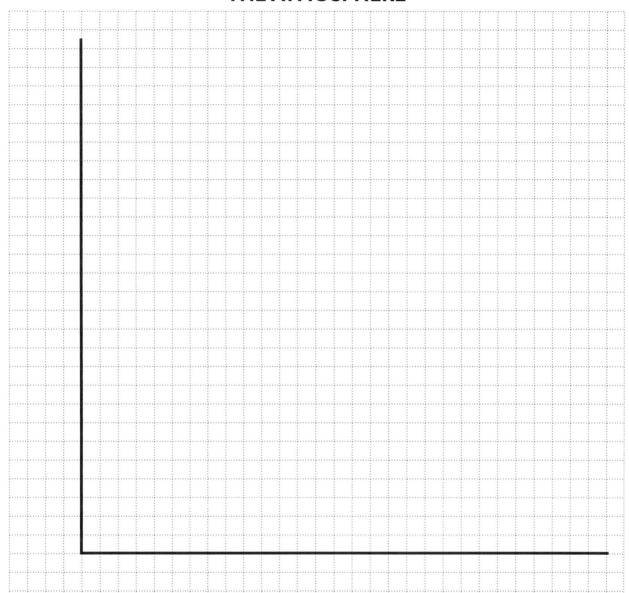

FIGURE C

Directions: (1) Set two Ehrlenmeyer flasks under a lit desk lamp or in the bright sun. (2) Light several matches and drop them into one of the flasks. (3) When the matches go out cap the flask with a plain rubber stopper and wait 3 minutes for the glass of the flask to cool. (4) Uncap the flask containing the matches and quickly insert one of the thermometer-stoppers provided by your instructor. (5) Cap the empty flask with the second thermometer-stopper. (6) Record the temperature of the thermometers every minute for 15 minutes and make 2 line graphs to show the rise in temperature in each of the flasks. Was there a difference? If so, what do you think caused the difference?

GENERAL SAFETY PRECAUTIONS

Exercise common sense precautions when working with matches.

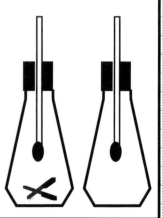

96

EA7 Review Quiz

Directions: Keep your eyes on your own work.
Read all directions and questions carefully.
THINK BEFORE YOU ANSWER!
Watch your spelling, be neat, and do the best you can.

CLASSWORK	(~40): _____
HOMEWORK	(~20): _____
CURRENT EVENT	(~10): _____
TEST	(~30): _____
TOTAL	(~100): _____

(A ≥ 90, B ≥ 80, C ≥ 70, D ≥ 60, F < 60)

LETTER GRADE: _____

TEACHER'S COMMENTS: _____

_____

THE ATMOSPHERE

TRUE–FALSE FILL-IN: If the statement is true, write the word TRUE. If the statement is false, change the underlined word to make the statement true. *20 points*

_____ 1. The earth is surrounded by a thin blanket of air we call the <u>geosphere</u>.

_____ 2. Air <u>is not</u> like all other forms of matter.

_____ 3. Air pressure at sea level is about 1,000,000 dynes per square centimeter or approximately <u>3.14</u> pounds per square inch.

_____ 4. One million dynes per square centimeter of atmospheric pressure is called one <u>stick</u>.

_____ 5. Evangelista Torricelli designed the first <u>thermometer</u> in 1643 to measure the force being exerted by the atmosphere.

_____ 6. Torricelli discovered that there was <u>less</u> atmospheric pressure surrounding him as he went to higher elevations.

_____ 7. The lowest layer of the atmosphere is called the <u>troposphere</u>.

_____ 8. The <u>stratosphere</u> absorbs much of the dangerous ultraviolet radiation that our planet receives from the sun.

_____ 9. The <u>mesosphere</u> cools slowly to an elevation of 85 kilometers from about 0°C to –90°C.

_____ 10. The atmosphere is made primarily of <u>nitrogen</u> gas.

MULTIPLE CHOICE: Choose the letter of the answer that best completes the sentence or answers the question. *10 points*

_____ 11. Which gas protects the earth's surface from ultraviolet radiation?

(A) nitrogen   (D) argon
(B) oxygen   (E) carbon dioxide
(C) ozone

_____ 12. Which is not a layer of the atmosphere?

(A) thermosphere   (D) troposphere
(B) mesosphere   (E) biosphere
(C) stratosphere

_____ 13. Where does weather occur?

(A) troposphere   (D) thermosphere
(B) stratosphere   (E) Van Allen Belts
(C) mesosphere

_____ 14. The field of magnetism and charged particles surrounding the earth is called the

_____.

(A) solar wind   (D) Aurora Australis
(B) Van Allen Belts (E) none of the above
(C) Aurora Borealis

_____ 15. Which is a major characteristic of air?

(A) it takes up space   (D) A and B, only
(B) it has mass   (E) A, B, and C are correct
(C) it exerts pressure

EARTH'S CLIMATE
AND MOST VIOLENT STORMS

TEACHER'S CLASSWORK AGENDA AND CONTENT NOTES

Classwork Agenda for the Week

1. Students will identify the world's major climatic zones.
2. Students will show how earth's rotation affects global wind patterns.
3. Students will identify global wind patterns.
4. Students will examine evidence that earth's climatic zones can change.

Content Notes for Lecture and Discussion

Aristotle (b. 384 B.C.; d. 322 B.C.) explained weather to be the result of an intermingling between the hot, dry "exhalations" of earth's interior and the cool, wet waters of the sea. Today, climate and weather are both considered the product of several factors: (1) the differential heating of the earth's curved surface by the sun, (2) the rotation of the earth, and (3) the topographical and biological features of the land. Solar radiation reaching the earth's surface is absorbed and reflected by the varying physical and chemical components of the planet's surface. Liquid water absorbs energy more slowly than solid matter but retains that energy for longer periods of time. Dark surfaces absorb more energy than lighter surfaces which reflect energy back into the atmosphere. Warm air expands and rises, pushed up by denser cold air that moves in to take its place. The winds created by this differential heating of the earth's surface encounter bodies of water where they pick up moisture and dry land features that dessicate them. In summary, the radiant energy of the solar wind is transformed into the kinetic energy of atmospheric motion. The kinetic energy of the atmosphere is dissipated by friction between air and surface features or converted to long wave heat energy that remains close to the planet's surface.

It was not until the middle of the 17th century that explorers began to see the world as a whole. Documentation of wind patterns across the face of the oceans allowed scientists to plot global wind patterns and attempt to explain them. The English lawyer and parttime astronomer **Edmond Halley** (b. 1656; d. 1742)—who discovered the comet that bears his name—suggested that the **trade winds** moving east to west across equatorial longitudes were the result of cold air arriving from polar and temperate zones to replace warm air rising at the equator. Halley's contemporary, **George Hadley** (b. 1685; d. 1768), suggested that the earth's rotation was the primary cause of equatorial winds. It was not until 1835, however, that the French physicist **Gaspard Gustave de Coriolis** (b. 1792; d. 1843) proved the point. As a result of the **Coriolis effect**, air masses in the Northern Hemisphere move in a clockwise direction. Air masses in the Southern Hemisphere move in a counterclockwise direction. Hadley later explained the movement of equatorial, temperate, and polar air masses by elucidating the idea of **convection cells**. Hadley proposed that warm air rising at the equator was replaced by colder air moving in from higher latitudes. The rising air at the equator moved to higher latitudes only to cool, become more dense, and sink. The **Hadley-three cell model** (see Illustration C in Lesson #3) included three distinct convection zones called **Hadley cells**. The equatorial and polar cells constituted air masses that rotated in the same direction. In between them, the temperate air masses were forced to rotate in the opposite direction by frictional contact with the two adjacent air masses.

German meteorologist **Wladmir Peter Köppen** (b. 1846; d. 1940) was the first to create a system of classification to describe **climatic zones**. He introduced his work in 1900 basing his classification system on measures of temperature and levels of rainfall tolerated by varying types of vegetation. Succeeding climatologists based their work on that of Köppen.

Today, climatologists worry that the atmosphere is changing. Geological and fossil evidence is filled with support for the idea that earth's climate is anything but static. In addition to the notion that continental drift has changed the topography of the planet which certainly altered climatic zones, there is evidence of numerous Ice Ages that have cooled the global temperature to bare minimums able to support life. The most widely accepted theory of the cause of the Ice Ages was suggested by the Yugoslavian physicist **Milutin Milankovich** (b. 1879; d. 1958). Milankovich considered the following facts: (1) the eccentricity of earth's elliptical orbit changes, placing the earth closer to and farther from the sun at periodic intervals; and (2) the earth wobbles on its axis altering its tilt in 21,000 year cycles (e.g., Polaris has not always been the "north star."). The result, Milankovich concluded, is a change in the reflective properties of the earth's absorbent surfaces. The atmosphere's inability to hold sufficient levels of solar radiation when the planet's tilt and proximity to the sun combine to minimize that absorption allows glaciation to proceed at an accelerated rate.

In Lesson #1, students will identify the world's major climatic zones and explain the significance of Hadley cells.

In Lesson #2, students will show how the earth's rotation affects global wind patterns and explain the Coriolis effect.

In Lesson #3, students will identify and map global wind patterns.

In Lesson #4, students will examine evidence that earth's climatic zones can change.

ANSWERS TO THE HOMEWORK PROBLEMS

Students should submit *a copy* of the partially completed weather chart they will use to complete the homework assignment in Unit 9: *Introduction to Meterology*. Urge them to use all information sources at their disposal to gather a sufficient amount of data (e.g., newspapers, Internet, television, and radio reports).

ANSWERS TO THE END-OF-THE-WEEK REVIEW QUIZ

1. climate	6. true	11. B	Hadley explained how air masses move
2. vary	7. front	12. C	across latitude lines in convection cells.
3. true	8. condenses	13. E	Coriolis explained the hemispheric
4. true	9. rods	14. F	rotation of global wind patterns.
5. true	10. cyclones, tornadoes, or hurricanes	15. A	
		16. D	

EA8 FACT SHEET

EARTH'S CLIMATE AND MOST VIOLENT STORMS

CLASSWORK AGENDA FOR THE WEEK

(1) Identify the world's major climatic zones.
(2) Show how earth's rotation affects global wind patterns.
(3) Identify global wind patterns.
(4) Examine evidence that earth's climatic zones can change.

Although the weather changes from day to day, average weather conditions over a large area of the earth's surface can remain constant over long periods of time. The long-term weather conditions over a large area of the earth are called the area's **climate**. Climate differs from one region to another because the amount of solar radiation reaching the earth's surface is different from one region to another. Places near the equator receive warm, direct rays from the sun. The North and South Poles receive cool, indirect rays. Areas in between the equator and the poles receive less energy than the equator but more energy than the poles. Climate is also determined by the topography of the earth's surface (e.g., mountains, valleys, and plains) and the type of vegetation that exists there.

There are five major climatic zones. **Tropical rain climates** are located in equatorial regions. **Polar climates** are found in the Arctic and Antarctic zones. **Dry climates** are found across latitudes in desert areas that receive little rain. **Warm** and **cool temperate rain climates** are also scattered across latitudes around the world.

Every day, warm air rises at the **equator** and flows toward the **poles**. As it moves toward the poles the air cools, sinks, and flows back toward the equator. This movement of air creates vertically circulating air masses that move across latitudes from the equator toward the poles and vice versa. The circulating air forms **convection cells** called **Hadley cells** after the English meteorologist **George Hadley** (b. 1685; d. 1768). As the earth rotates on its axis from west to east air masses tend to flow in the opposite direction from east to west. This westerly flow of air causes wind patterns to circulate in a clockwise direction in the Northern Hemisphere and a counterclockwise direction in the Southern Hemisphere. This phenomenon was discovered in 1835 by the French physicist **Gaspard Gustave de Coriolis** (b. 1792; d. 1843) and is called the **Coriolis effect**.

A flowing air mass moves along a border called a **front**. **Cold fronts** are heavier and usually wedge under **warm fronts**. As warm air rises, the water vapor in the air cools and condenses into droplets. Clouds form and **precipitation** (i.e., snow, sleet, hail, rain) can occur. Moving air masses frequently result in fierce weather conditions. **Thunderstorms** are violent storms that develop on hot humid days. During a thunderstorm, static electricity can be discharged violently from rumbling clouds to the ground or vice versa. **Lightning rods** are used to conduct electric charges into the earth without damage to buildings and people. An **occluded front** occurs when a cold front moves faster than a warm front and overtakes it. This condition can result in the rapid upward movement of spinning air called a **cyclone**. **Tornadoes** are small violent cyclones that occur over land. Hurricanes are large cyclones that develop over the oceans.

Homework Directions

Begin collecting data about the weather conditions in your area from newspapers, the Internet, television, and radio reports. Set up a chart to keep a record of the following information for the next ten days: (1) daily high and low temperatures, (2) relative humidity readings, (3) barometric pressure readings, (4) wind speed and direction, (5) general sky conditions (e.g., clear skies, partly cloudy skies, etc.). Be prepared to turn in *a copy* of your chart this week containing the first three days of information. *Continue collecting data through next week.* Your homework assignment in Unit 9: *Introduction to Meteorology* will require you to make a graph of this and additional data.

Assignment due: _____

_____   _____   ____/____/____
Student's Signature   Parent's Signature   Date

EA8 Lesson #1

EARTH'S CLIMATE AND MOST VIOLENT STORMS

Work Date: ____/____/____

LESSON OBJECTIVE

Students will identify the world's major climatic zones.

Classroom Activities

On Your Mark!

Ask students to explain the difference between weather and climate. After some brief discussion write the definitions on the board and have students copy each definition onto Journal Sheet #1. **Weather** refers to local atmospheric conditions with regard to temperature measurements, wind speed and direction measurements, humidity and cloud formations. **Climate** refers to the weather conditions that exist over a large area of the earth for a long period of time. List the three important factors that cause the climate to be different from one region to another: (1) differential heating of the earth's curved surface by the sun, (2) the rotation of the earth, and (3) the differential absorption of solar rays by the variety of rocks, soils, plant and animal life that exist in different parts of the world.

Get Set!

Draw Illustration A on the board and have students copy it onto Journal Sheet #1. Explain that solar radiation from the sun strikes the earth at varying angles due to the earth's curved surface. Ask them to imagine a bright flashlight being shined in their eyes. If they tilted their head back, then the strong direct light would become weaker and indirect.

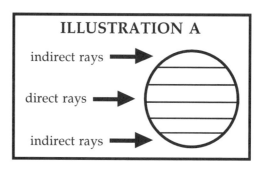

ILLUSTRATION A

indirect rays →

direct rays →

indirect rays →

Go!

Have students perform the activity described in Figure A on Journal Sheet #1. When they are finished with this activity have them compare the map of "The World's Major Climatic Zones" to a political map of the earth. Have them write the names of two countries in each major climatic zone.

Materials

dark construction paper or index cards, dissecting needle, cardboard box, Journal Sheet #1, paper clips, flashlight, political map of the world

EA8 JOURNAL SHEET #1

EARTH'S CLIMATE AND MOST VIOLENT STORMS

FIGURE A

Directions: (1) Use the template shown below to punch holes in a piece of construction paper or index card with a dissecting needle. Place the paper on a cardboard box to avoid injury and damage to underlying surfaces. (2) Roll this JOURNAL SHEET into a cylinder and secure it with paperclips so that the curve of the cylinder goes from the North to South Poles. (3) Aim a lit flashlight at the cylinder from a distance of about 20-30 centimeters. (4) Place the construction paper or index card 2-4 centimeters from the surface of the cylinder. (5) Move the card up and down while noting the distance between the points of light shining on the cylinder. (6) What happens to the distance between the pinpoints of light as they move across the surface of the cylinder? How does the curve of the earth affect the amount of sunlight reaching its surface?

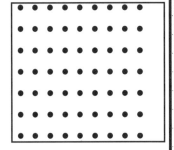

THE WORLD'S MAJOR CLIMATIC ZONES

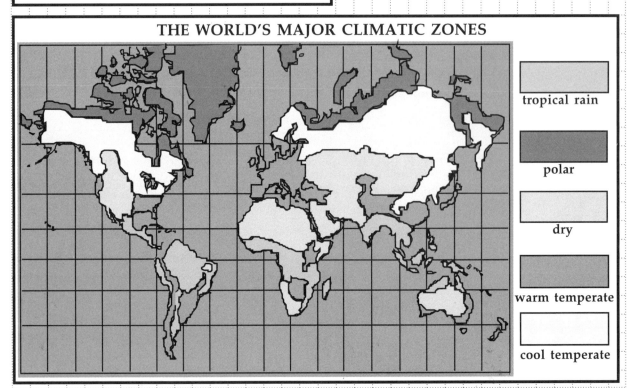

tropical rain

polar

dry

warm temperate

cool temperate

EARTH'S CLIMATE AND MOST VIOLENT STORMS

Work Date: _____/_____/_____

LESSON OBJECTIVE

Students will show how earth's rotation affects global wind patterns.

Classroom Activities

On Your Mark!

Begin the lecture by telling students that seafaring explorers of the 17th, 18th, and 19th centuries—when merchant vessels were powered by the wind—took advantage of the winds that blew westward across the oceans at the equator. Seafarers called these winds the **trade winds** because they were so essential to setting up profitable trade routes. No one could explain why the trade winds were so reliable until the French physicist **Gaspard Gustave de Coriolis** (b. 1792; d. 1843) tackled the problem.

Get Set!

Perform the following demonstration to explain Coriolis's reasoning. (1) Tie a 1-hole rubber stopper to a meter-long length of string. (2) Spin the stopper in a large circle over your head allowing the free end of the string to dangle through your fist. (3) Ask students to observe what happens to the rotational speed of the stopper as you pull it in. Pull on the string with your free hand to reduce the length of the string. Students will observe that the speed of the stopper increases as the distance between it and your stationary hand decreases. As the earth rotates on its axis, the atmosphere rides along with it. However, air masses moving away from the equator are also moving closer to the center of the earth along its rotational axis (e.g., a straight line drawn from the surface to the earth's center). The velocity of the air therefore increases just as the rubber stopper did when you "pulled it in." That is why objects appear to be deflected to the right when moving north away from the equator and to the left when moving south away from the equator. The opposite is the case for air masses moving from the poles toward the equator. The velocity of these moving air masses decreases, deflecting them westward as the world spins under them. This phenomenon is called the **Coriolis effect**. Draw Illustration B on the board. Have students copy your illustration onto Journal Sheet #2.

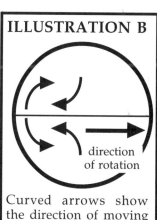

ILLUSTRATION B

direction of rotation

Curved arrows show the direction of moving air masses toward and away from the equator as a result of the Coriolis effect.

Go!

Have students perform the activity described in Figure B on Journal Sheet #2.

Materials

1-hole rubber stopper, string, paper cups, paper plate, food coloring, pencils

Name: _____ Period:_____ Date: ___/___/___

EA8 JOURNAL SHEET #2

EARTH'S CLIMATE AND MOST VIOLENT STORMS

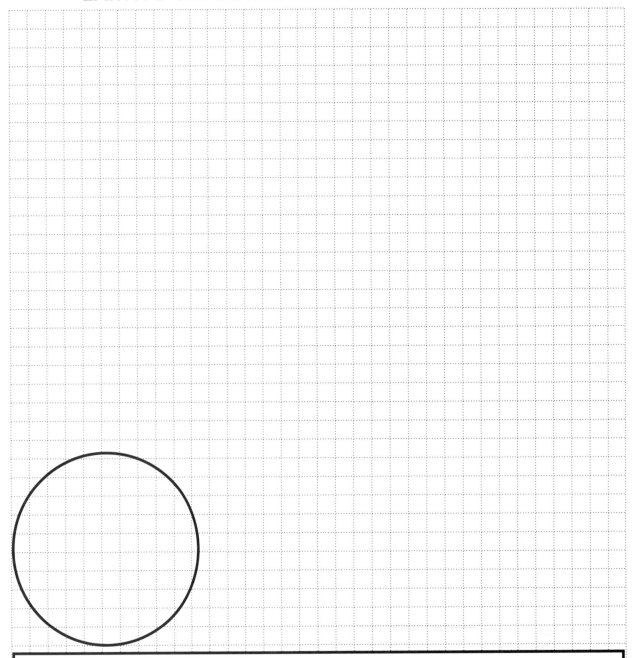

FIGURE B

Directions: (1) Punch a hole through the center of a paper plate and the bottom of a paper cup with a pencil. (2) Place the assembly on a table as shown. (3) Hold the pencil securely and rotate the paper plate until it begins to spin freely around the pencil. (4) Put several drops of food coloring in the plate making a circle of drops about 1-2 centimeters around the pencil. (5) Begin rotating the plate counter-clockwise. (6) Draw your observations of the lines created by the food coloring in the above circle.

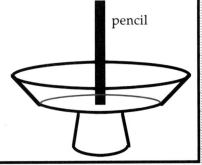

pencil

106

EA8 Lesson #3

EARTH'S CLIMATE AND MOST VIOLENT STORMS

Work Date: ____/____/____

LESSON OBJECTIVE

Students will identify global wind patterns.

Classroom Activities

On Your Mark!

Begin the lecture with a review of the **Coriolis effect**. Explain that while the earth's rotation is responsible for the deflection of large air masses across longitude lines as the air moves away from the equator, it is the differential heating of the earth's surface that causes the winds to start. Point out that different materials absorb and reflect different amounts of heat. Dark surfaces absorb a lot of solar radiation and convert that energy to long wave infrared radiation. Ask students to report what they have observed happening over hot asphalt in a parking lot during summer. Have they ever seen ripples of heat rising over a barbeque? Mention English meteorologist **George Hadley** (b. 1685; d. 1768) who explained how air masses circulate vertically in the atmosphere to produce three convection cells surrounding the planet in each hemisphere. Draw Illustration C to help students visualize **Hadley's three-cell model**. Draw the illustration that appears on Journal Sheet #3 if a transparency is unavailable and assist students in labelling the major "wind belts" of the world created by the Coriolis effect and Hadley convection cells. The **northeast** and **southeast trade winds** blow east to west from about 10° to 30° N. and S. latitudes. Between them are the **doldrums** where equatorial winds are calm. From about 30° to 60° N. and S. latitudes flow the **stormy westerlies**. **Polar easterlies** circulate around the poles from east to west.

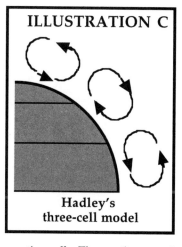

ILLUSTRATION C

Hadley's three-cell model

Get Set!

Draw Illustration D to show how the borders of large air masses can interact to produce violent storms called cyclones. Explain that tornadoes are devastating cyclones with winds up to 500 kilometers per hour. Tornadoes occur over land. The most active tornado zone in the world—"Tornado Alley"—is located in the central plain states of the United States. Hurricanes with winds up to 200 kph are much larger but more slowly moving cyclones. Hurricanes originate over the ocean. Have students copy your illustration onto Journal Sheet #3.

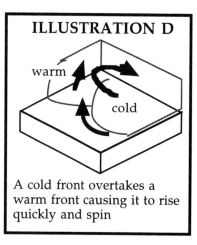

ILLUSTRATION D

warm

cold

A cold front overtakes a warm front causing it to rise quickly and spin

Go

Have students perform the activity described in Figure C on Journal Sheet #3. <u>Caution students not to inhale the smoke created by the burnt matches. Warn them to use common sense precautions</u> and to follow directions.

Materials

scissors, shoe boxes, plastic straws, matches, birthday candles, plastic bottle caps, petri dish, ice or dry ice (if available)

107

EA8 JOURNAL SHEET #3

EARTH'S CLIMATE AND MOST VIOLENT STORMS

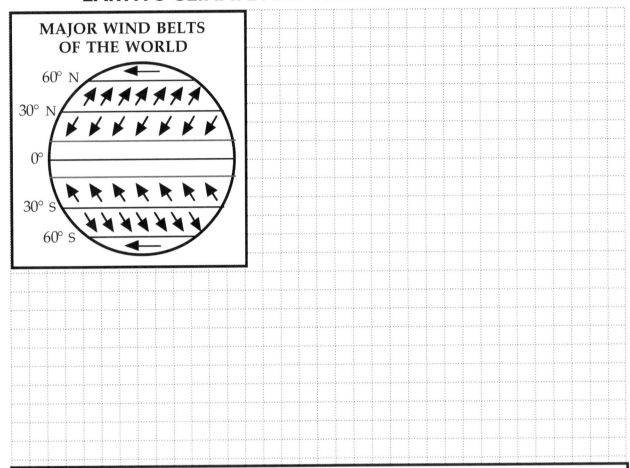

MAJOR WIND BELTS OF THE WORLD

60° N
30° N
0°
30° S
60° S

FIGURE C

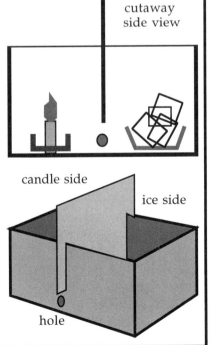

cutaway side view

candle side

ice side

hole

Directions: (1) Punch a small hole with a pencil near the bottom middle of a shoe box. (2) Cut slots in the lid of the shoe box and slide the lid over the edges of the box as shown to divide the box into two compartments. Leave room at the bottom of the box (e.g., 1-2 cm) to allow the free exchange of air between sides. (3) Melt several drops of wax off the bottom of a birthday candle allowing the wax to drip into the bottom of a plastic bottle cap. Quickly place the candle into the melted wax securing the candle to the inside of the cap as the wax cools and hardens. Place the cap in the center of one side of the box. (4) Place some ice in a small petri dish on the other side of the box. Use dry ice if available and ignore steps #5 through #8 and simply report on the direction that the carbon dioxide vapors flow. (5) Stand a straw on a table and light a match. (6) Drop the lit match -- flame side up -- down the straw. Light and drop another match down the straw in similar fashion. (7) Plug both ends of the straw with your fingers and insert one end into the hole in the shoe box. (8) Ask a classmate to watch what is happening from above and report the direction that smoke flows as you gently blow into the other end of the straw without touching the straw to your lips. A light breath will force the smoke from the matches into the box.

EA8 Lesson #4

EARTH'S CLIMATE AND MOST VIOLENT STORMS

Work Date: ____/____/____

LESSON OBJECTIVE

Students will examine evidence that earth's climatic zones can change.

Classroom Activities

On Your Mark!

Begin lecture by pointing out that there is substantial evidence to indicate that earth's climate has not been static over the ages. The annual growth rings of trees suggest that winters may be harsh in otherwise warm regions for decades or longer, shrinking the growth of the rings. Continental drift—having changed the position of the continents—has no doubt changed the climatic zones of the world by altering wind patterns over land and sea. Glacial ages have led to the extinction of millions of species of plants and animals only to pass as global climate changed again. It is also probable that asteroid impacts have thrown enough dust into the atmosphere to block the light of the sun altering the climate in a myriad of ways.

Get Set!

Draw Illustration E as a possible explanation for how Ice Ages occur and have students copy the illustration on Journal Sheet #4. Introduce the Yugoslavian physicist **Milutin Milankovich** (b. 1879; d. 1958) and his theory discussed in the Teacher's Classwork Agenda and Content Notes. **Axial wobble** alters the tilt of the planet every 21,000 years while causing a precession of the equinoxes. That is, the seasons shift in time across the calendar as the perihelion and aphelion (e.g., earth's closest and farthest distances from the sun) shift along earth's "wobbily" orbital path. At one extreme, winter in the northern hemisphere occurs at aphelion. At present, winter occurs at perihelion. The combination of maximal tilt at aphelion could be the cause of the Ice Ages.

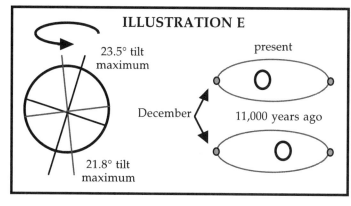

ILLUSTRATION E

23.5° tilt maximum

21.8° tilt maximum

December

present

11,000 years ago

Go!

Allow students sufficient time to brainstorm scenarios that might cause the earth's climate to change. Urge them to use all of the information they have learned in this and preceding units. Have them write a "sci-fi" plot in which one or more of the possible climatological catastrophies occur.

Materials

Journal Sheet #4

EA8 JOURNAL SHEET #4

EARTH'S CLIMATE AND MOST VIOLENT STORMS

_____
(Story Title)

EA8 Review Quiz

Directions: Keep your eyes on your own work.
Read all directions and questions carefully.
THINK BEFORE YOU ANSWER!
Watch your spelling, be neat, and do the best you can.

CLASSWORK	(~40): _____
HOMEWORK	(~20): _____
CURRENT EVENT	(~10): _____
TEST	(~30): _____
TOTAL	(~100): _____

(A ≥ 90, B ≥ 80, C ≥ 70, D ≥ 60, F < 60)

LETTER GRADE: _____

TEACHER'S COMMENTS: _____
_____

EARTH'S CLIMATE AND MOST VIOLENT STORMS

TRUE–FALSE FILL-IN: If the statement is true, write the word TRUE. If the statement is false, change the underlined word to make the statement true. *10 points*

_____ 1. Long term weather conditions over a large area of the earth is called the area's <u>weather</u>.

_____ 2. The varying amounts of solar radiation that reach the earth's surface cause climate to <u>remain the same</u> around the planet.

_____ 3. The North Pole receives <u>cool, indirect</u> solar rays.

_____ 4. The equator receives <u>warm, direct</u> solar rays.

_____ 5. Vertically circulating air masses that flow back and forth from the equator to the poles are called <u>Hadley cells</u>.

_____ 6. <u>The Coriolis effect</u> cause(s) air masses to flow from east to west across the planet as it rotates on its axis.

_____ 7. A flowing air mass flows across a border called a <u>cloud</u>.

_____ 8. As warm air rises, water vapor cools and <u>evaporates</u> to form clouds.

_____ 9. Lightning <u>bolts</u> conduct electricity in the air to the ground without damaging buildings.

_____ 10. Cold air pressing warm air into an occluded front can cause <u>calm weather</u>.

MATCHING: Choose the letter of the phrase that best describes each major climatic zone. *12 points*

_____ 11. tropical rain   (A) cold severe winters
_____ 12. arid   (B) rain most or all of the year
_____ 13. semi-arid   (C) dry desert
_____ 14. warm temperate   (D) cold all year
_____ 15. cool temperate   (E) sparse rain
_____ 16. polar   (F) rainy winter

EA8 Review Quiz *(cont'd)*

ESSAY: In a few short sentences explain the contributions of scientists George Hadley and Gustave de Coriolis. *8 points*

_____

_____

_____

_____

_____

_____

_____

_____

_____

_____

_____

EA9 INTRODUCTION TO METEOROLOGY

Teacher's Classwork Agenda and Content Notes

Classwork Agenda for the Week

1. Students will demonstrate the hydrological cycle and how clouds form.
2. Students will demonstrate the uses of the barometer and anemometer.
3. Students will demonstrate the uses of the thermometer and hygrometer.
4. Students will plot a weather map.

Content Notes for Lecture and Discussion

The work of early scientists like **Edmond Halley** (b. 1656; d. 1742), **George Hadley** (b. 1685; d. 1768), and **Gaspard Gustave de Coriolis** (b. 1792; d. 1843) helped to discover the cause of the trade winds and to elucidate the circulation patterns of large air masses due to the **Coriolis effect** and the impact of solar radiation on the earth's surface. But the behavior of everyday weather, from balmy breezes to devastating storms, could not be understood until large-scale coordinated observations of atmospheric conditions could be made. Around the middle of the 17th century, The Royal Society of London organized a small network of weather stations that attempted to communicate by horse and carriage information about local weather conditions over towns and villages outside London. Of course, the speed of the communication was—more often than not—outpaced by the changing weather condition itself. In the late 18th century, the Mannheim Meteorological Society set up an extensive network of stations that permitted scientists to draw the first weather charts of continental Europe. Ten years after the devastating East Coast Hurricane of 1821, American meteorologist **William Redfield** (b. 1789; d. 1857) identified the tropical origin of hurricanes. Redfield also established the fact that hurricanes were rotating masses of turbulent air. Immediately following the invention of the telegraph in 1838 by **Samuel F. B. Morse** (b. 1791; d. 1872), American physicist **Joseph Henry** (b. 1797; d. 1878) was instrumental in setting up a sophisticated network of telegraph stations that evolved into the United States Weather Bureau. By the start of the Civil War, there were more than 500 telegraph stations reporting to the Smithsonian Institute in Washington, D.C., providing information used to compose daily weather maps for every American newspaper.

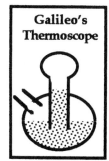

Galileo's Thermoscope

Instruments used to measure the various characteristics of the atmosphere have a long history dating back to the first **absorption hygrometer** or "hygroscope" to measure moisture in the air. The hygroscope was invented by the German Cardinal **Nicholas de Cusa** in the 15th century (b. 1401; d. 1464). De Cusa's invention was a simple balance with a piece of damp wool suspended at one end and balanced by weights at the other. The Cardinal noted that the wool became heavier as the air grew in dampness. Modern hygrometers measure **relative humidity** by comparing the air temperature of a "dry bulb" thermometer to the temperature of a "wet bulb" thermometer. The less water vapor in the air, the greater the evaporation and cooling of the wet bulb. The greater the cooling of the wet bulb thermometer, the greater the difference between the two thermometers. **Galileo Galilei** (b. 1564; d. 1642) designed the first "thermoscope" in 1593. The thermoscope shown at right was used to measure air temperature. Cool air caused the air in the sealed bulb to contract, permitting atmospheric pressure to push the liquid in the lower chamber up the tube. Warm air had the opposite effect. By

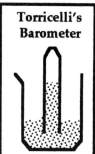

Torricelli's Barometer

1718, Dutch physicist **Gabriel Daniel Fahrenheit** (b. 1686; d. 1736) had introduced the **Fahrenheit scale** to give units of measure to temperature readings. Swedish scientist **Anders Celsius** (b. 1701; d. 1744) introduced the **centigrade scale** in 1742. The Italian scientist **Evangelista Torricelli** (b. 1608; d. 1647) invented the barometer shown on the previous page in 1643 by inverting a mercury-filled glass tube sealed at one end into a bowl of mercury. Taking his invention to higher elevations caused the mercury in the tube to fall and leave an empty space at the top of the tube. Torricelli had not only demonstrated a method for measuring atmospheric pressure but became infamous for proving that a vacuum could exist. **Aristotle** (b. 384 B.C.; d. 322 B.C.) had wrongly deduced that a vacuum could not exist in nature. The invention and use of other instruments—the **anemometer** to measure wind speed, **rain gauges** to measure the amount of rainfill, **weather balloons**, and **weather satellites**—have added to the store of data used to interpret atmospheric motion. The French admiral and meteorologist **Francis Beaufort** (b. 1774; d. 1857) established the first velocity scale used to measure the visible effects of the wind. Journal Sheet #4 of this unit contains a summary of Beaufort's Scale. Perhaps the most accomplished meteorologist was the Norwegian scientist **Vilhelm Firman Koren Bjerknes** (b. 1862; d. 1951). It was Bjerknes who coined the term "front" to describe the boundaries between moving air masses, drawing an analogy between those air masses and opposing armies in battle. Bjerknes's theories concerning the motion of polar air masses and the hydrodynamic models of the oceans and atmosphere form the basis of modern meteorology. The first of the modern "weathermen," Bjerknes demonstrated how weather forecasts could be made on a statistical basis, lending the power of mathematics to this newly developing science.

In Lesson #1, students will demonstrate the hydrological cycle and how clouds form.

In Lesson #2, students will demonstrate the uses of the barometer and construct a vane anemometer.

In Lesson #3, students will demonstrate the uses of the thermometer and show how a hygrometer is used to determine relative humidity.

In Lesson #4, students will plot a weather map, and explain the causes of violent storms.

ANSWERS TO THE HOMEWORK PROBLEMS

Students should use the data gathered in Unit 8: *Earth's Climate and Most Violent Storms* along with additional data from this week's homework assignment to complete tables and graphs showing how the weather has behaved in the last few weeks.

ANSWERS TO THE END-OF-THE-WEEK REVIEW QUIZ

1. weather	6. pressure	11. D	16. E
2. from the sky	7. true	12. B	17. F
3. sun	8. warm, dry	13. G	
4. true	9. cool, wet	14. A	
5. Communication	10. warm, wet	15. C	

ESSAY: The 3 conditions required to form a cloud are (1) water vapor, (2) dust particles, and (3) a decrease in air pressure and temperature.

EA9 FACT SHEET

INTRODUCTION TO METEOROLOGY

CLASSWORK AGENDA FOR THE WEEK

(1)   Demonstrate the hydrological cycle and how clouds form.
(2)   Demonstrate the uses of the barometer and anemometer.
(3)   Demonstrate the uses of the thermometer and hygrometer
(4)   Plot a weather map.

The Greek philosopher **Aristotle** (b. 384 B.C.; d. 322 B.C.) thought that weather was the result of hot, dry gases from the earth's interior rising and mixing with the cold, moist air of the oceans. Today, scientists know that weather is the product of many factors. They know that the engine that drives earth's weather is the sun. Solar radiation heats the oceans and many surfaces of the earth at different rates. The differential heating of the earth's waters and landforms causes the wind. It also changes the amount of water vapor in the atmosphere by evaporating water from oceans, lakes, and streams. Differential heating of earth's surfaces by the sun also changes the pressure and density of the air in any given place. The study of the weather is called **meteorology**. The term "meteorology" comes from the Greek word "meteoron" which means "from the sky." In Aristotle's day, the study of meteorology included the study of meteors: particles of rock that invade earth's atmosphere from outer space.

Scientists who study the weather are called **meteorologists**. Meteorologists use a number of instruments to measure local atmospheric conditions. But they must know what the weather is in many different areas before they can predict how it will change from region to region. **Communication** is an essential part of weather forecasting. Before the invention of the telegraph by the American inventor **Samuel F. B. Morse** (b. 1791; d. 1872) in 1838 forecasting the weather was impossible. Morse's invention allowed scientists to make maps of weather conditions over a wide range of areas and to begin predicting how air masses might move.

Meteorologists use several essential tools to measure atmospheric conditions. A **weather vane** is used to indicate the direction that the wind is blowing. It is a simple device: as simple as a tree bending in the wind. An **anemometer** measures the speed of the wind. A **thermometer** measures air temperature. And, a **barometer** measures air pressure. A **hygrometer** measures the amount of moisture in the air. With these tools, meteorologists determine the conditions of the air in their local area and communicate that information to a central weather station. The centralized weather station compiles data from many meteorologists in the field and draws a weather map. Today, **weather satellites** are of great value in helping meteorologists to forecast the weather. From high above earth's surface, orbiting satellites provide a "bird's eye view" of our planet. Satellites take still and motion pictures of moving cloud formations and can even measure the temperature and amount of water vapor in the air below.

The study of meteorology has allowed scientists to identify a number of major air masses that move across the planet's surface every year. By studying these patterns of air movement, they hope to better predict the weather for all of us.

EA9 Fact Sheet *(cont'd)*

Homework Directions

Continue collecting data about the weather conditions in your area from newspapers, the Internet, television, and radio reports. Continue charting your record of the following information for the next several days: (1) daily high and low temperatures, (2) relative humidity readings, (3) barometric pressure readings, (4) wind speed and direction, (5) general sky conditions (e.g., clear skies, partly cloudy skies, etc.). Including the data you collected in the last unit on *Earth's Climate and Most Violent Storms,* graph your weather readings to show how they have changed in past weeks.

Assignment due: _____

_____   _____   ___/___/___
Student's Signature   Parent's Signature   Date

EA9 Lesson #1

INTRODUCTION TO METEOROLOGY

Work Date: ____/____/____

LESSON OBJECTIVE

Students will demonstrate the hydrological cycle and how clouds form.

Classroom Activities

On Your Mark!

Prepare the apparatus shown in Figure A on Journal Sheet #1 before the start of class. Prepare a solution of "sea water" by mixing salt with sand or muddy pond water. Set the mixture aside in a large jar or beaker.

Begin class discussion with a review of the changing phases of matter (e.g., solid, liquid, and gas) and how water is recycled and purified during the **hydrological cycle**. Explain that sea water is a mixture, a solution made of many substances that can be separated by warming. Draw Illustration A on the board and have students copy the illustration onto Journal Sheet #1.

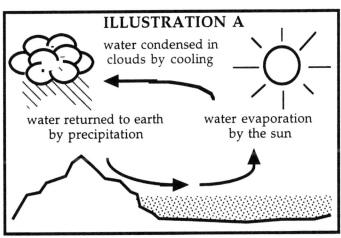

ILLUSTRATION A

water condensed in clouds by cooling

water returned to earth by precipitation

water evaporation by the sun

Get Set!

Explain that water molecules cohere more easily to one another when there is a foreign substance present to bring them closer together. Microscopic dust particles in the air serve as the "nuclei" at the center of rain droplets around which water molecules can gather and stick together. As water vapor rises into the atmosphere air pressure and temperature decrease, reducing the momentum of individual water molecules and making them more susceptible to the electrostatic Van der Waals forces that hold liquid water molecules together. Clouds are formed in the presence of (1) water vapor, (2) dust particles, and (3) reduced air pressure. Introduce students to the terminology used to describe the most common types of clouds: alto = high; cumulus = fluffy; stratus = straight; nimbus = rain; cirrus = whispy "mare's tail." These terms are used in combination to describe a variety of cloud formations (e.g., cumulonimbus, nimbostratus, stratocumulus, altocumulus, etc.)

Go!

Give students ample time to set up and perform the activities described in Figure A and Figure B on Journal Sheet #1.

Materials

ringstand and clamps, Ehrlenmeyer flasks, single-holed rubber stoppers, glass tubing, small beakers, Bunsen burners, plain water and "sea water" mixture, matches, large balloons, ice, pie pans

117

EA9 JOURNAL SHEET #1

INTRODUCTION TO METEOROLOGY

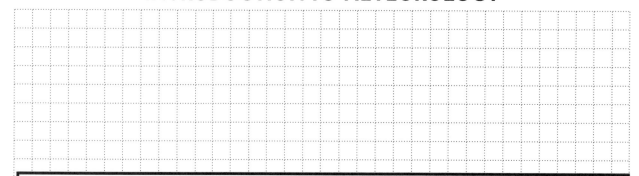

FIGURE A

Directions: (1) Pour into an Ehrlenmeyer flask 100 milliliters of the "sea water" mixture given to you by your instructor. (2) Pour about 50 ml of the same solution into a small beaker. (3) Place the flask on the ringstand and secure it with a clamp so that it cannot be toppled. (4) Place the beaker next to the ring stand. (5) Insert the rubber stopper holding the glass tubing snugly into the flask. DO NOT HOLD OR MANIPULATE THE GLASS TUBING. HANDLE THE RUBBER STOPPER ONLY. Make sure the other end of the glass tubing is positioned over the beaker. (6) Turn on the Bunsen burner. (7) Record what you observe for the next ten minutes and explain your observations. What is the color and consistency of the evaporated water compared to the "sea water"? What process allows the pure water to return to the "sea"?

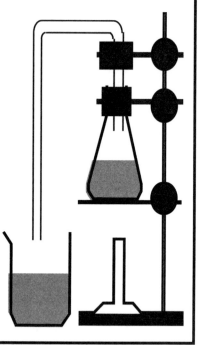

GENERAL SAFETY PRECAUTIONS

Be sure you are familiar with the proper use of a Bunsen burner. Wear goggles to protect your skin and eyes from being burned by SCALDING HOT STEAM. Do not touch any part of the equipment without heat-resistant gloves or tongs. Clean up when the apparatus is cool.

FIGURE B

Directions: (1) Place two small beakers into a bowl of ice to act as "cloud chambers" for the purpose of this experiment. (2) Pour about 25 ml of water into each beaker. (3) Cut two balloons so that they can be wrapped snugly over the rim of each beaker. Allow the balloons to sag slightly so that they can be pulled and stretched to increase the volume of the air inside the cloud chamber. (4) Stretch one of the balloons over one of the beakers. (5) Light several matches and toss them into the second uncovered beaker allowing smoke to drift around inside the beaker. (6) Cover the second beaker with the second balloon before all of the smoke dust escapes. (7) Give the beakers a moment to cool, then pull up on the balloons covering each beaker to expand the volume of the air and decrease the pressure in each chamber. (8) Explain your observations and list the three conditions required to form a cloud.

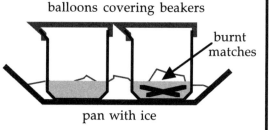

balloons covering beakers

burnt matches

pan with ice

GENERAL SAFETY PRECAUTIONS

Wear goggles and use common sense precautions when working with matches.

INTRODUCTION TO METEOROLOGY

Work Date: ____/____/____

LESSON OBJECTIVE

Students will demonstrate the uses of the barometer and anemometer.

Classroom Activities

On Your Mark!

Referring to the Teacher's Classwork Agenda and Content Notes in this unit and to Lesson #1 in Unit 7: *The Atmosphere,* begin discussion with a review of how the atmosphere exerts pressure. Remind students of the discoveries made by the Italian scientist **Evangelista Torricelli** (b. 1608; d. 1647). Repeat the demonstrations shown in Illustrations B and C in Unit 7. Point out that 1,000 millibars of air pressure is enough to hold up a column of liquid mercury about 760 millimeters high. If a vacuum pump is available perform the following demonstration shown in Illustration B: (1) Construct a Torricelli barometer by inserting a water-filled test tube into a beaker of water. (2) Place the barometer into the vacuum chamber, seal it with sealing grease, and turn it on. (3) Instruct students to watch the water level in the inverted tube as the air is "sucked out" of the vacuum chamber. As atmospheric pressure is reduced in the chamber gravity pulls down the liquid in the tube.

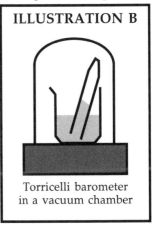

ILLUSTRATION B

Torricelli barometer
in a vacuum chamber

Get Set!

Explain that water is a satisfactory analogy for air since both air and water are fluids that exert pressure. Point out that the device they are going to use in the activity described in Figure C on Journal Sheet #2 can be used to measure both ocean depth and, with some minor adjustments, "atmospheric depth" (e.g., how deep one is below the surface of the atmosphere).

Remind students that winds are caused by the differential heating of the earth's surface. As warm air expands and rises cooler air pushes in to take its place. Discuss the uses of a **weather vane** (e.g., an airport wind sock) and have students discuss the places where they have seen weather vanes. Explain that an **anemometer** like the one shown in Figure D on Journal Sheet #2 is really any device used to measure the speed of the wind. Modern anemometers are connected to gauges that count the turns of a spinning rod attached to the cups.

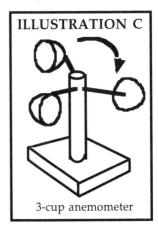

ILLUSTRATION C

3-cup anemometer

Go!

Give students ample time to set up and perform the activity described in Figure C on Journal Sheet #2.

Materials

ringstand and clamps, large buckets, rubber tubing, thistle tubes or funnels, balloons, meter stick, glass U-tube prepprepared by bending a piece of glass over a Bunsen burner, index cards, pencils

EA9 JOURNAL SHEET #2

INTRODUCTION TO METEOROLOGY

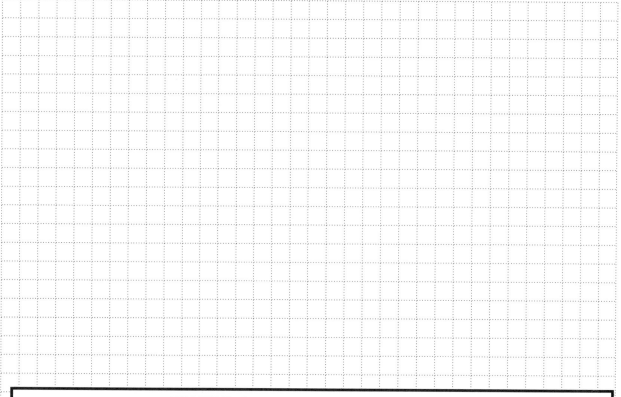

FIGURE C

Directions: (1) Fill a large bucket with water. (2) Clamp a metric ruler to a ringstand beside the bucket. (3) Cut a balloon and wrap it over the mouth of a thistle tube or funnel. (4) Fill half-way with colored water the U-tube provided by your instructor. USE CAUTION IN HANDLING THE U-TUBE. IT IS MADE OF FRAGILE GLASS THAT CAN BREAK. (5) Attach one end of a length of rubber tubing to the funnel and the other end to the glass U-tube as shown. (6) Attach the U-tube to a clamp placed in front of the metric ruler. (7) Submerge the end of the funnel into the bucket to varying depths and record what happens to the level of the water in the U-tube.

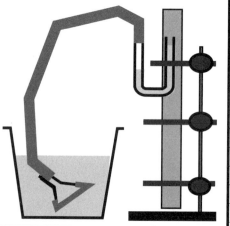

FIGURE D

Directions: (1) Refer to the illustration at left to construct a simple vane anemometer. (2) Use a protractor to draw a semicircle with divisions on one index card and fold the card over the edge of a table. (3) Attach another index card with paper clips to a straw or pencil so that the "vane" swings freely. (4) Lay the straw or pencil across the center of the first index card as shown and blow softly in the direction shown by the arrow. Explain how a more sophisticated device using this design could be used to measure wind speed more accurately.

INTRODUCTION TO METEOROLOGY

Work Date: ____/____/____

LESSON OBJECTIVE

Students will demonstrate the uses of the thermometer and hygrometer.

Classroom Activities

On Your Mark!

Begin discussion with a review of the three most commonly used temperature scales. The **Fahrenheit** and **Celsius** (e.g., **centigrade**) **scales** are the most well known. The **Kelvin scale** is used primarily by scientists as an "absolute" measure of temperature related to the amount of heat energy existing in a given object. The temperature of outer space is about 3 Kelvin above **absolute zero** (e.g., the temperature at which all molecular motion would theoretically cease). Explain that all temperature scales are arbitrary scales derived from measurements of the behavior of water when it boils or freezes at 1,000 millibars pressure.

Get Set!

Perform the following demonstration to illustrate the meaning of **relative humidity**: (1) Weigh a dry sponge on a balance and record it on the board. (2) Submerge the sponge in water until it is totally saturated and weigh it again. Record the result and have students calculate the amount of water absorbed by the sponge (e.g., the total weight after submersion less the weight of the dry sponge). (3) Inform students that the sponge is "100% saturated." That is, it cannot hold any more water. It is at 100% humidity! (4) Squeeze some of the water out of the sponge and weigh it again. Record the result and ask students to calculate the amount of water in the sponge now. (5) Divide the weight of the water after squeezing by the weight of the water before squeezing and multiply by 100 to give a percent. (6) Explain that the percent answer is the "relative humidity" of the sponge after squeezing or the amount of water it held after squeezing compared to the maximum amount it could hold when it is saturated.

Go!

Give students ample time to set up and perform the activity described in Figure F on Journal Sheet #3.

Materials

thermometers, cotton or paper towels, books, water, sponges, balance

EA9 JOURNAL SHEET #3

INTRODUCTION TO METEOROLOGY

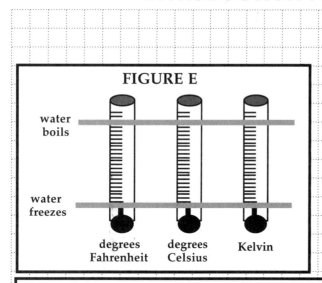

FIGURE E

water boils

water freezes

degrees Fahrenheit   degrees Celsius   Kelvin

RELATIVE HUMIDITY CHART

dry bulb temperature (°F)

wet bulb temperature (°F)	60	65	70	75	80	85	90	95
45	26	13	3					
50	48	30	19	9	3			
55	73	52	36	24	15	8	3	
60		66	55	40	29	20	13	10
65			77	53	44	33	24	15
70				78	61	47	36	27
75					79	63	45	39
80						81	65	52
85							81	66
90								82

The air holds more water when it is warmer than when it is cold. And, when the air is fully saturated wet and dry bulb readings are the same regardless of temperature. As the air becomes drier, evaporation cools the wet bulb and lowers its temperature. The greater the difference in temperatures between the wet and dry bulb the lower the relative humidity.

FIGURE F

<u>Directions</u>: (1) Wrap the bulbs of two thermometers in cotton or paper towel. (2) Secure the thermometers in between the pages of a textbook as shown. (3) Place a towel underneath the set up and soak the wrapping around the first thermometer with water.
Leave the wrapping around the other thermometer dry. (4) Allow the thermometers to sit for 10 minutes; then read the temperatures. (5) Compare the readings to the temperature scales shown on the RELATIVE HUMIDITY CHART. What is the relative humidity of your classroom?

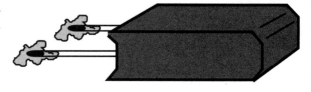

EA9 Lesson #4

INTRODUCTION TO METEOROLOGY

Work Date: ___/___/___

LESSON OBJECTIVE

Students will plot a weather map.

Classroom Activities

On Your Mark!

Begin with a discussion of the importance of **communication** in the development of accurate forecasting. Use the information in the Teacher's Classwork Agenda and Content Notes to introduce students to the scientists who made modern forecasting possible. Draw a quick sketch of the continental United States on the board and explain that there are six major air masses that flow across the Northern Hemisphere. Polar air masses flowing down from the North Pole carry cold air. If the air moves down over the continent (e.g., **polar continental**), then the air tends to be dry. If the air moves down over the oceans (e.g., **polar maritime**), then it tends to be wet. Tropical air masses flowing up from the equator carry warm air. If the air moves up over the continent (e.g., **tropical continental**), then the air tends to be dry. If the air moves up over the oceans (e.g., **tropical maritime**), then it tends to be wet.

Get Set!

Refer to the map in Figure G on Journal Sheet #4. Explain that each dot represents a barometric pressure reading from a single weather station. Modern weather stations containing all of the instruments needed to measure atmospheric conditions are placed on floating **buoys** at sea or in convenient locations across the continent. Measurements are sent by radio or telephone cable to the United States Weather Bureau where computers compile and analyze the data. Maps are drawn and sent to local television, radio, and newspaper stations where the information can be reported to the public. The map shown in Figure G is an **isobar map**. Isobar lines indicate regions of equal air pressure.

Go!

Give students ample time to complete the activity described in Figure G on Journal Sheet #4. They should conclude that the map was probably drawn during the summer or fall since the high pressure zones are located in the south. These high pressure zones will tend to push north against the lower pressure zones. Both tropical maritime and tropical continental air masses have a higher pressure than the polar continental air mass suggesting that the continent is in for some "rough weather" when the multiple fronts meet.

Materials

Journal Sheet #4

EA9 JOURNAL SHEET #4

INTRODUCTION TO METEOROLOGY

BEAUFORT SCALE AND WIND EFFECTS

Beaufort number	wind condition	wind speed (kph)
0	calm	<1
1	light air	1–5
2	light breeze	6–11
3	gentle breeze	12–19
4	moderate breeze	20–28
5	fresh breeze	29–38
6	strong breeze	39–49
7	near gale	50–61
8	gale	62–74
9	strong gale	75–88
10	storm	89–102
11	violent storm	103–114
12–17	hurricane	117 or above

FIGURE G

Directions: (1) Connect dots of equal pressure. (2) Identify the polar continental, polar maritime, tropical continental, and tropical maritime air masses that flow over the continental United States. (3) Write a paragraph forecast of the weather that will take place over the continent. Will the wind flow generally from East-to-West or West-to-East? South-to-North or North-to-South? Does this map indicate a period of summer or winter? Explain your reasoning.

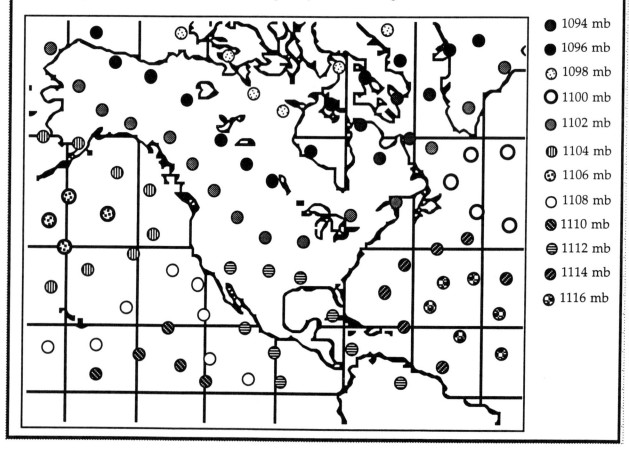

- 1094 mb
- 1096 mb
- 1098 mb
- 1100 mb
- 1102 mb
- 1104 mb
- 1106 mb
- 1108 mb
- 1110 mb
- 1112 mb
- 1114 mb
- 1116 mb

EA9 Review Quiz

Directions: Keep your eyes on your own work.
Read all directions and questions carefully.
THINK BEFORE YOU ANSWER!
Watch your spelling, be neat, and do the best you can.

CLASSWORK   (~40): _____
HOMEWORK   (~20): _____
CURRENT EVENT   (~10): _____
TEST   (~30): _____

TOTAL   (~100): _____
(A ≥ 90, B ≥ 80, C ≥ 70, D ≥ 60, F < 60)

LETTER GRADE: _____

TEACHER'S COMMENTS: _____

_____

INTRODUCTION TO METEOROLOGY

TRUE–FALSE FILL-IN: If the statement is true, write the word TRUE. If the statement is false, change the underlined word to make the statement true. *10 points*

_____ 1. Short term atmospheric conditions over a small area of the earth is called <u>climate</u>.

_____ 2. The Greek word "meteoron" means <u>a big rock</u>.

_____ 3. The engine that drives the earth's weather is the <u>earth's core</u>.

_____ 4. Scientists who study the weather are called <u>meteorologists</u>.

_____ 5. <u>A satellite</u> is an essential part of weather forecasting.

_____ 6. Isobar lines on a weather map show areas of equal air <u>temperature</u>.

_____ 7. The polar continental is a <u>cool, dry</u> air mass.

_____ 8. The tropical continental is a <u>cool, dry</u> air mass.

_____ 9. The polar maritime is a <u>cool, dry</u> air mass.

_____10. The tropical maritime is a <u>cool, dry</u> air mass.

ESSAY: List the three conditions necessary to form a cloud. *6 points*

(1) _____

_____

(2) _____

_____

(3) _____

_____

EA9 Review Quiz (cont'd)

MATCHING: Choose the letter of the word or phrase at right that best describes the function performed by the meteorological instrument at left. *14 points*

_____ 11. weather vane   (A) measures wind speed

_____ 12. thermometer   (B) measures air temperature

_____ 13. barometer   (C) measures air moisture

_____ 14. anemometer   (D) measures wind direction

_____ 15. hygrometer   (E) reports large scale cloud movement

_____ 16. satellites   (F) helps meteorologists to make maps

_____ 17. communication networks   (G) measures air pressure

MAPPING THE HEAVENS

TEACHER'S CLASSWORK AGENDA AND CONTENT NOTES

Classwork Agenda for the Week

1. Students will use a compass and astrolabe to map objects in the sky.

2. Students will compare and contrast refracting and reflecting telescopes.

3. Students will use the changing positions of familiar constellations to prove that the earth revolves around the sun.

4. Students will use parallax to determine the distance to faraway objects.

Content Notes for Lecture and Discussion

Astronomy began in prehistoric times proving of great value to farmers and seafarers. The former required the invention of an accurate calendar to assist them in preparing for the planting and harvesting of crops at appropriate times of the year. The latter would have had difficulty navigating home when sailing out of sight of land had they not had a rudimentary knowledge of the stars. The Babylonians and Egyptians were accomplished astronomers, having kept accurate records of the motions of the sun, moon, stars, and visible planets. The Egyptians attributed the motion of celestial objects to metaphysical phenomena and associated them with ancient religious lore. By 800 B.C. the Egyptians had developed a form of theoretical astrology to explain natural and social events on earth, considering out of the ordinary celestial events as omens: good and bad. Many of their beliefs were adopted by early Greeks who devised the familiar horoscopes associated with the religion. Unlike modern **astronomy**, a scientific discipline, astrology is a religion based on unsubstantiated belief. Although astrologers can be sophisticated in their understanding of planetary motion and the interpretation of their presumed effects on people's lives, an astrologer's efforts lack the experimental component necessary to withstand the rigors of empirical investigation. The ancient Chinese concentrated their efforts in the documenting of uncommon celestial events such as the appearance of comets, eclipses, and novae which they also attributed to metaphysical causes.

The Greeks' fascination with mathematics—particularly geometry—was the beginning of the systematic study of astronomy. The works of **Plato** (b. 427 B.C.; d. 347 B.C.), **Aristotle** (b. 384 B.C.; d. 322 B.C.), and **Euclid** (b. 330 B.C.; d. 260 B.C.) emphasized the regular patterns traced out by the sun, moon, stars, and planets. They sought to explain these movements according to the precise formulations of mathematical axioms and theorems. The invention of the **astrolabe** by **Hipparchus** (b. 190 B.C.; d. 120 B.C.) in 150 B.C. furthered their accumulation of precise data.

The mathematics of Ancient Greece was adopted by future astronomers who made the motions of the planets predictable if not explicable. With the work of the Egyptian astronomer **Claudius Ptolemaeus** (b. 100; d. 170) and the Polish astronomer **Nicolaus Copernicus** (b. 1473; d. 1543) a picture of the universe began to emerge. Ptolemy's **Geocentric Theory** was the most widely accepted picture of the solar system for over a thousand years. He placed the earth at the center of the universe with the sun, moon, stars, and planets moving around us. While the sun, moon, and stars appeared to make nearly perfect circles around the earth, the planets wandered back-and-forth, changing direction and speed, in patterns which Ptolemy called **epicycles**. The word "planet" comes from the Greek term "plané" meaning "wanderer." Nicolaus Copernicus explained the illusion of the epicycles, basing his work largely on the observations of others. He put the sun at the center of the system with all of the planets—including earth—moving around it. Copernicus's **Heliocentric Theory**, introduced in his work *On the Revolutions of the Heavenly Spheres*, was not published until after his death and was condemned by the Roman Catholic Church as a heretical document. It remained on the church's roll of "forbidden books" until 1835.

EA10 Content Notes *(cont'd)*

Even the accomplished Danish astronomer **Tycho Brahe** (b. 1546; d. 1601)—whose extensive works allowed the German mathematician **Johannes Kepler** (b. 1571; d. 1630) to prove that the planets moved around the sun in elliptical orbits—was a "geocentrist" to his death. In the middle of the 17th century, the English mathematician and scientist **Sir Isaac Newton** (b. 1642; d 1727) derived his **Laws of Motion** and a **Universal Law of Gravitation**. Newton's Law of Gravity explained the motions of the planets in one thoroughly encompassing idea. And although it was later replaced by the **Theory of General Relativity**, proposed by the American physicist, **Albert Einstein** (b. 1879; d. 1955), Newton's Law served to put American astronauts **Neil Alden Armstrong** (b. 1930) and **Buzz Aldrin** (b. 1930) on the moon in 1969.

Following the invention of the telescope by **Hans Lippershey** (b. 1570; d. 1619) in 1608 and its extensive use by **Galileo Galilei** (b 1564; d. 1642)—who used it to popularize Copernicus's Geocentric Theory—other inventions have expanded the study of astronomy. With the theoretical and technological advances of the 20th century (e.g., the invention of radio telescopes that see beyond the visible spectrum) astronomy has widened its horizons. The study of astronomy has joined the study of cosmology in attempting to elucidate the origin and fate of the universe.

In Lesson #1, students will explain the difference between the Geocentric and Heliocentric Theories of the solar system and use a compass and astrolabe to map objects in the sky.

In Lesson #2, students will compare and contrast refracting and reflecting telescopes.

In Lesson #3, students will use the changing positions of familiar constellations to prove that the earth revolves around the sun.

In Lesson #4, students will use parallax to determine the distance to faraway objects.

ANSWERS TO THE HOMEWORK PROBLEMS

Students' graphs should look like the one below. Considering the times at which the movements of this "unidentified flying object" were observed one might reasonably conclude that it is indeed an alien U.F.O. Point out, however, that the distance to the object is unknown and those measurements need to be considered as well as other evidence before jumping to any conclusion. A helicopter that is close by might maneuver in such a manner and disappear by turning off its running lights.

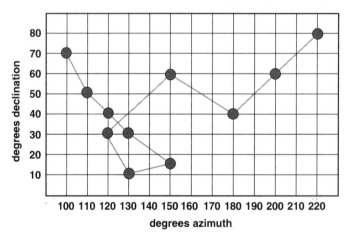

ANSWERS TO THE END-OF-THE-WEEK REVIEW QUIZ

1. astronomy
2. constellations
3. calendars
4. true
5. true
6. true
7. telescope
8. true
9. true
10. true
11. E
12. A
13. C
14. B
15. D
16. F

ILLUSTRATION: The telescopes must be placed in their appropriate boxes and contain all of the essential elements shown in the diagrams on Journal Sheet #2.

EA10 FACT SHEET

MAPPING THE HEAVENS

CLASSWORK AGENDA FOR THE WEEK

(1) Use a compass and astrolabe to map objects in the sky.
(2) Compare and contrast refracting and reflecting telescopes.
(3) Use the changing positions of familiar constellations to prove that the earth revolves around the sun.
(4) Use parallax to determine the distance to faraway objects.

The study of the stars is called **astronomy** and it is a very ancient science. Without the aid of sophisticated tools (e.g., like a telescope) most anyone can see the sun, moon, stars, and planets Venus, Mars, Jupiter, and Saturn, if they know where and when to look. Ancient people recognized arrangements of stars which they called **constellations**; and they realized that the movement of the stars and planets went along with the change of the seasons. They invented tools like the astrolabe and built monuments like the giant blocks at Stonehenge to help them schedule the planting and harvesting of crops. Planting too early or too late in the season could mean the difference between eating or starving. So, farmers made calendars based on the movements of **celestial objects**. The first calendar was a lunar calendar based on the changing phases of the moon. Modern calendars are based on earth's revolution around the sun and the rising and setting of constellations during the year.

Although the ancients did not understand the forces that moved the sun, moon, stars, and planets, they saw the importance of keeping a record of these movements. They used the mathematics of geometry—their knowledge of circles and the angles in triangles—to help them draw detailed maps of the sky. The word "geometry" means "to measure the earth." Early observers drew maps showing the location of objects across (e.g., **azimuth**) and above (e.g., **declination**) the horizon. One of the first great astronomers was a Greek named **Hipparchus** (b. 190 B.C.; d. 120 B.C.). He is given credit for the invention of the astrolabe in 150 B.C. An **astrolabe** is a device used to measure the altitude or angle of objects above the horizon. Using his astrolabe, Hipparchus improved ancient methods of locating places on earth using longitude and latitude lines. In addition, he made a permanent record of nearly 1,000 stars by mapping their fixed location in the sky.

Ancient people watched the sun rise in the East and set in the West every day of their lives as you do. At night, they saw the stars move in circles around the sky "revolving" around the fixed North Star. The planets Venus, Mars, Jupiter, and Saturn were seen wandering around the night sky in roller coaster loops that seemed to change their direction and speed. Their conclusion was that the earth was the center of the universe and that all celestial objects moved around us. This theory of the universe was called the **Geocentric Theory** after the Egyptian astronomer **Claudius Ptolemaeus** (b. 100; d. 170). More than a millenium later, in 1513, the Polish astronomer **Nicolaus Copernicus** (b. 1473; d. 1543) proposed another theory. Although his theory was condemned and his books outlawed, an increasing number of astronomers began to accept it. Copernicus's theory was called the Heliocentric Theory and says that all of the planets including earth revolve around the sun. The term "helio" means "sun." In 1608, a Dutch lensmaker named **Hans Lippershey** (b. 1570; d. 1619) designed the first **refracting telescope**. His invention was used by the Italian astronomer **Galileo Galilei** (b. 1564; d. 1642) to prove that Nicolaus Copernicus was correct.

EA10 Fact Sheet (cont'd)

In 1619, the German mathematician and astronomer **Johannes Kepler** (b. 1571; d. 1630) formulated a series of laws to describe planetary motion around the sun. **Kepler's laws** have since been used to make predictions about the paths of comets and planets, giving a complete description of the past, present, and future movement of these objects. The famous English mathematician and philosopher **Sir Isaac Newton** (b. 1642; d. 1727) used Kepler's laws to derive his **Universal Law of Gravitation** in 1684. Newton is also given credit for having invented the first reflecting telescope after which all modern telescopes are designed.

Homework Directions

Use graph paper to plot the motion of the object described in Table A. What conclusions can you draw about this object based on its movements across the night sky?

TABLE A		
time	degrees azimuth	degrees declination
8:50 pm	100	70
8:51 pm	110	50
8:52 pm	120	40
8:53 pm	130	30
8:54 pm	150	15
8:55 pm	130	10
8:56 pm	120	30
8:57 pm	150	60
8:58 pm	180	40
8:59 pm	200	60
9:00 pm	220	80
9:01 pm	object gone	

Assignment due: _____

_____   _____   ___/___/___
Student's Signature   Parent's Signature   Date

MAPPING THE HEAVENS

Work Date: ____/____/____

LESSON OBJECTIVE

Students will use a compass and astrolabe to map objects in the sky.

Classroom Activities

On Your Mark!

Give students a brief introduction to the history of **astronomy**, defining astronomy as the scientific study of the stars, using the information in the Teacher's Classwork Agenda and Content Notes. Briefly discuss the difference between astronomy and astrology which is a religion based on unsubstantiated belief (e.g., faith). Draw Illustration A on the board and discuss the difference between the **Ptolemaic Geocentric** and the **Copernican Heliocentric Theories** of the solar system. Inform students that both of these theories were derived after careful "mapping" of the skies or "celestial sphere." Point out that the term "celestial" means "heavenly." Define the **celestial sphere** as the imaginary ceiling of the universe which, like the ceiling of a planetarium, appears to have many fixed objects stuck to it (e.g., faraway stars and galaxies) as well as other celestial objects that appear to move against that background. Define the term **azimuth** (e.g., given in Lesson #2, Unit 1, *Mapping the Earth*) as the distance or angle across the horizon measured from due North. Define **declination** as the angle above the horizon (e.g., 0°) to the **zenith** directly above one's head (e.g., 90°).

Get Set!

Draw Illustration B on the board to show how the planet Mars appears to move in the night sky (e.g., from position a' to b' to c' to d') when plotted on a sky chart. Give each position an azimuth and declination reading so that students can plot each position on Journal Sheet #1. By Copernicus's reasoning, however, Mars did not actually "wander in epicycles" across the sky, but instead moved in a larger, slower circle than the earth around the sun. Just as a car you might pass on the highway appears to be moving backward, the planet Mars appears to move backward as the earth passes ahead of it in a faster orbit closer to the sun.

Go!

Give students ample time to perform activity described in Figure A on Journal Sheet #1.

ILLUSTRATION A
Ptolemy's Geocentric Theory

planet in epicycle · sun · moon · earth

Copernicus's Heliocentric Theory

planet · moon · earth · sun

ILLUSTRATION B

azimuth

declination

b' c'
d'   a'

d c b a

Mars moves around the sun more slowly and at a greater distance from the sun than earth (e.g., positions a, b, c, and d). When projected onto a sky chart from the viewpoint of earth, however, it appears to alter speed and direction in the sky (e.g., positions a', b', c', and d').

Materials

nuts or brass weights, string, scissors, hole punch, plastic straws, compass and **azimuth indicator** from Lesson #2, Unit 1: *Mapping the Earth*

EA10 JOURNAL SHEET #1

MAPPING THE HEAVENS

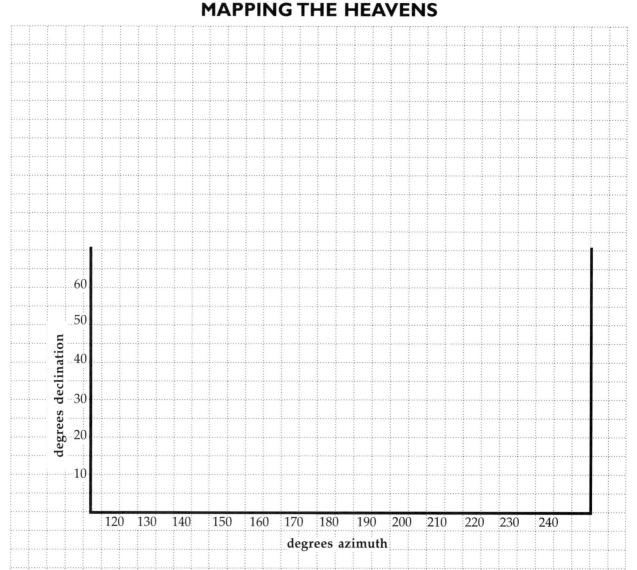

degrees declination

60
50
40
30
20
10

120 130 140 150 160 170 180 190 200 210 220 230 240

degrees azimuth

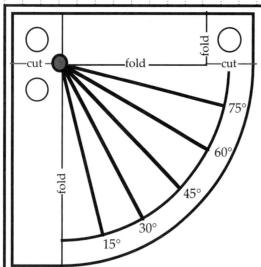

cut
fold
fold
cut
fold

75°

60°

45°

30°

15°

fold

FIGURE A

<u>Directions</u>: (1) Trace the astrolabe template to the left and cut it out. (2) Glue the tracing to a piece of construction paper, then punch out the holes, cut and fold as indicated so the astrolabe looks like the drawing at right. (3) Hang a weight tied to a piece of string as shown. (4) Insert a straw to use as a "viewer." (5) Measure the angle of any object above the horizon by noting the angle indicated by the vertically suspended string.

MAPPING THE HEAVENS

Work Date: ____/____/____

LESSON OBJECTIVE

Students will compare and contrast refracting and reflecting telescopes.

Classroom Activities

On Your Mark!

Begin class with a brief introduction to the work of the German mathematician-astronomer **Johannes Kepler** (b. 1571; d. 1630). Draw Illustration C on the board and have students copy your drawing onto Journal Sheet #2. Use the illustration to explain two of Kepler's laws of planetary motion: (1) Planets orbit the sun in an **ellipse** and not a circle. Explain the difference between an ellipse and a circle. While a circle has one center called a "focus," an ellipse has two "foci" that cause the figure to take on an "oval" shape. (2) The radius vector of each planet sweeps out equal areas in equal times. Point out the shaded areas at perihelion and aphelion in the orbit of the planet indicated. The areas of both shaded zones are equal. The planet will sweep out both these areas in the same amount of time. This of course means that when the planet is closer to the sun its velocity increases. Comets have highly elliptical orbits. Point out that Kepler did much of his work before the telescope was invented by **Hans Lippershey** (b. 1570; d. 1619) in 1608.

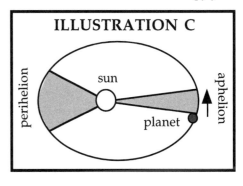

ILLUSTRATION C

perihelion · sun · aphelion · planet

Get Set!

Use Illustration D to assist students in labeling the parts of the **refracting** and **reflecting telescopes** used by **Galileo Galilei** (b. 1564; d. 1642) and **Sir Isaac Newton** (b. 1642; d. 1727), respectively. Point out that the reflecting telescope has the advantage of being easier to construct and handle because mirrors can be made of polished metal. This makes it possible to build larger and larger telescopes that can see farther and farther away.

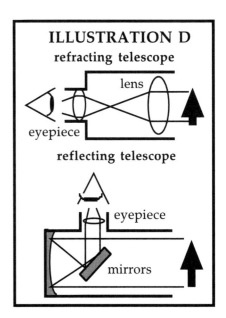

ILLUSTRATION D

refracting telescope

lens

eyepiece

reflecting telescope

eyepiece

mirrors

Go!

Give students ample time to perform the activities described in Figure B on Journal Sheet #2. Students will note a larger, inverted image of the candle when viewed through the correctly positioned magnifying lenses. They should also be able to produce a "floating virtual image" above the soup spoon when the flashlight is placed within an inch of the spoon. Students will be tempted to reach out and touch the image. As soon as their finger gets in the way, the image disappears.

Materials

mirrors, magnifying lenses, construction paper, scissors, flashlight

EA10 JOURNAL SHEET #2

MAPPING THE HEAVENS

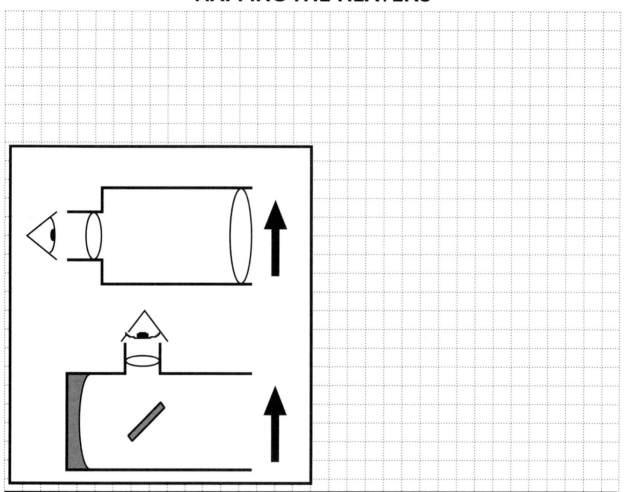

FIGURE B

<u>Directions:</u> (1) Melt the bottom of a birthday candle into a plastic bottle cap, then secure the candle to the melted wax. (2) Place the candle on a table and darken the room. (3) Light the candle and move the first magnifying glass back and forth until the lit candle comes into focus. (4) Move a piece of looseleaf paper between you and the lens until an image of the candle appears on the paper. (5) Put the paper down, pick up the second magnifying glass, and place it in the same position as the image when it appeared on the paper. You may have to move the lens slightly back and forth to bring the image of the candle back into view. (6) Record your observations. What is the position of the candle? Does it appear larger or smaller than the candle itself? How do you think the mountains <u>on the top</u> half of the moon looked to Galileo when he first saw them through his **refracting telescope**?

<u>Directions:</u> (1) Cut a piece of cardboard or dark construction paper to fit over the head of a flashlight. (2) Cut the block letter "T" into the cardboard and tape it to the flashlight. (3) Place a large, cleanly polished soup spoon "bowl up" on the table. (4) Darken the room. (5) Turn on the flashlight and bring it close to the spoon until an image of the "T" appears. (6) Record your observations. Where does the image appear? Does it appear flat against the back of the spoon or can you make it "float in thin air" above the spoon? This "floating image" is called a virtual image. In a **reflecting telescope**, the virtual image is reflected off a mirror into the eyepiece.

GENERAL SAFETY PRECAUTIONS
Use common sense safety guidelines when dealing with matches, lit candles, and hot wax!

MAPPING THE HEAVENS

Work Date: ____/____/____

LESSON OBJECTIVE

Students will use the changing positions of familiar constellations to prove that the earth revolves around the sun.

Classroom Activities

On Your Mark!

Draw Illustration E to give students a visual impression of earth and the celestial sphere. Explain that the celestial sphere is the "ceiling" of the universe against which all of the distant stars and galaxies appear "fixed."

Get Set!

Draw Illustration F on the board and have students copy it on Journal Sheet #3, listing the 12 constellations of the "zodiac." Explain that as the earth revolves around the sun, a different constellation rises above the horizon at sunset every month. These constellations were used by ancient astronomers to create a reliable calendar. When the earth is in the position shown the constellation "Leo" rises at sunset as the sun disappears over the horizon. This arrangement of the stars in relation to the earth and the sun gave Copernicus one of his first clues that the sun was the center of the known universe.

ILLUSTRATION E

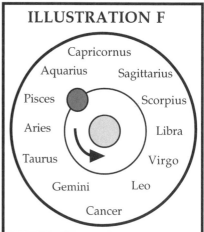

ILLUSTRATION F

Go!

Give students ample time to perform the activity described in Figure C on Journal Sheet #3.

Materials

construction paper, balls of different size

EA10 JOURNAL SHEET #3

MAPPING THE HEAVENS

CONSTELLATION KEY

Capricornus = goat
Sagittarius = archer
Scorpius = scorpion
Libra = weight scales
Virgo = maiden
Leo = lion
Cancer = crab
Gemini = twins
Taurus = bull
Aries = ram
Pisces = fish
Aquarius = water carrier

FIGURE C

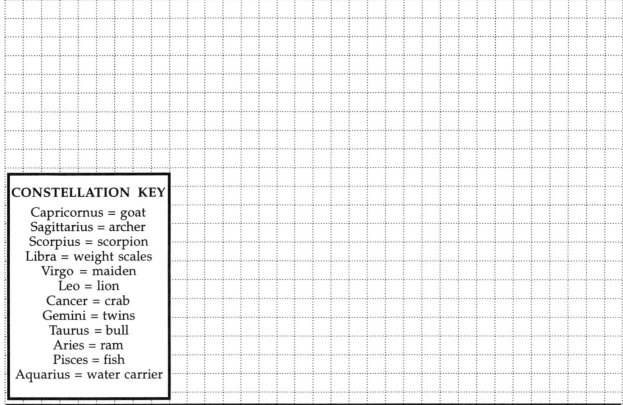

Capricornus   Sagittarius   Scorpius   Libra   Virgo   Leo

Aquarius   Pisces   Aries   Taurus   Gemini   Cancer

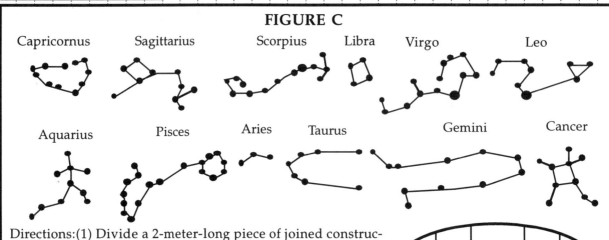

Directions:(1) Divide a 2-meter-long piece of joined construction or butcher paper into 12 boxes. (2) In each box, draw an enlarged replica of each of the above constellations in clockwise order beginning with Capricornus and ending with Aquarius. Use the CONSTELLATION KEY to surround each constellation with a cartoon that illustrates the meaning of its name. (3) Attach the ends of the paper so that Capricornus and Aquarius come together as shown. (4) Place the larger of 2 balls (e.g., representing the sun) at the center of the "zodiac" and move the smaller ball (e.g., representing the earth) around the larger ball. (5) Explain why Copernicus considered this arrangement of the earth, moon, and stars as proof that the sun and not the earth was the center of the known universe.

MAPPING THE HEAVENS

Work Date: ____/____/____

LESSON OBJECTIVE

Students will use parallax to determine the distance to faraway objects.

Classroom Activities

On Your Mark!

Have students perform the following demonstration: (1) Instruct them to hold an index finger vertically in front of their nose. (2) Tell them to close one eye and look at the finger. (3) Tell them to open that eye and close the other eye to observe the finger. (4) Ask: "Does the finger appear to shift position?" Answer: Yes. (5) Instruct them to move their finger out to arm's length and repeat steps #2 and #3. (6) Ask: "Does the finger appear to shift more or less with respect to the wall than it did when it was closer to their nose?" Answer: The finger appears to shift less when it is farther away. <u>Conclusion</u>: The farther an object is from an observer the less it will appear to shift position when viewed from two different angles. That is parallax! Define **parallax** as the apparent shift in the position of a distant object when viewed from different angles. Have them copy this definition onto Journal Sheet #4.

Get Set!

Draw Illustration G on the board to show how astronomers estimate the distance to faraway stars. Point out that a knowledge of triangles allows scientists to find the length of any side of a "right triangle" if the angle and one other side of the triangle is known. The Parallax Chart on Journal Sheet #4 gives the tangent value of the angle created by the rubber stopper, the earth, and sun in the activity described in Figure D. Scientists will plot the position of a star against the "fixed stars" of the celestial sphere and wait six months to plot the position of that same star. Stars within several

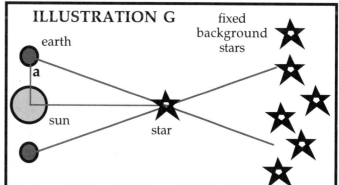

ILLUSTRATION G

earth — sun — star — fixed background stars

The tangent of angle "a" is equal to the distance to the star divided by the distance of the earth to the sun (e.g., the tangent of any angle in a right triangle is the ratio of the side opposite the angle to the side adjacent to the angle). Multiplying the known distance to the sun by the tangent of angle "a" gives a reliable estimate of the distance to the star.

hunded light-years of earth appear to shift against the background when viewed from these two orbital positions, making an estimate of their distance possible by this method.

Go!

Give students ample time to perform the activity described in Figure D on Journal Sheet #4.

Materials

rubber stoppers

EA10 JOURNAL SHEET #4

MAPPING THE HEAVENS

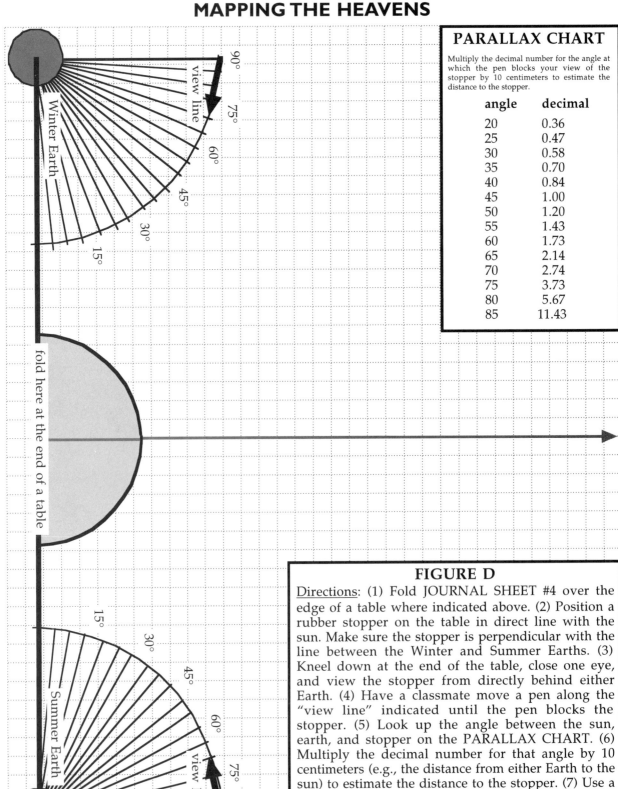

PARALLAX CHART

Multiply the decimal number for the angle at which the pen blocks your view of the stopper by 10 centimeters to estimate the distance to the stopper.

angle	decimal
20	0.36
25	0.47
30	0.58
35	0.70
40	0.84
45	1.00
50	1.20
55	1.43
60	1.73
65	2.14
70	2.74
75	3.73
80	5.67
85	11.43

FIGURE D

Directions: (1) Fold JOURNAL SHEET #4 over the edge of a table where indicated above. (2) Position a rubber stopper on the table in direct line with the sun. Make sure the stopper is perpendicular with the line between the Winter and Summer Earths. (3) Kneel down at the end of the table, close one eye, and view the stopper from directly behind either Earth. (4) Have a classmate move a pen along the "view line" indicated until the pen blocks the stopper. (5) Look up the angle between the sun, earth, and stopper on the PARALLAX CHART. (6) Multiply the decimal number for that angle by 10 centimeters (e.g., the distance from either Earth to the sun) to estimate the distance to the stopper. (7) Use a ruler to measure the distance from the center of the sun to the stopper to verify the accuracy of your observation.

EA10 REVIEW QUIZ

Directions: Keep your eyes on your own work.
Read all directions and questions carefully.
THINK BEFORE YOU ANSWER!
Watch your spelling, be neat, and do the best you can.

CLASSWORK	(~40): _____
HOMEWORK	(~20): _____
CURRENT EVENT	(~10): _____
TEST	(~30): _____
TOTAL	(~100): _____

(A ≥ 90, B ≥ 80, C ≥ 70, D ≥ 60, F < 60)

LETTER GRADE: _____

TEACHER'S COMMENTS: _____
_____

MAPPING THE HEAVENS

TRUE–FALSE FILL-IN: If the statement is true, write the word TRUE. If the statement is false, change the underlined word to make the statement true. *10 points*

_____ 1. The scientific study of the stars is called <u>astrology</u>.

_____ 2. Recognized arrangements of stars in the sky are called <u>correlations</u>.

_____ 3. The first <u>telescopes</u> were used to schedule the planting and harvesting of crops.

_____ 4. The distance around the horizon from due North is called <u>azimuth</u>.

_____ 5. The altitude or angle above the horizon is called <u>declination</u>.

_____ 6. A(n) <u>astrolabe</u> is a device used to measure the altitude or angle of an object above the horizon.

_____ 7. A(n) <u>telegraph</u> is a device used to see objects that are far away.

_____ 8. The farther an object is from an observer the <u>less</u> it appears to shift against the background when viewed from different angles.

_____ 9. According to Ptolemy's theory, the <u>earth</u> is the center of the solar system.

_____ 10. According to Copernicus's theory, the <u>sun</u> is the center of the solar system.

EA10 Review Quiz *(cont'd)*

MATCHING: Choose the letter of the scientific contribution at right with the scientist who made it at left. *12 points*

_____ 11. Hipparchus
_____ 12. Ptolemy
_____ 13. Copernicus
_____ 14. Lippershey
_____ 15. Kepler
_____ 16. Newton

(A) developed Geocentric Theory
(B) invented the refracting telescope
(C) developed the Heliocentric Theory
(D) showed correctly how planets move
(E) invented the astrolabe
(F) discovered the Law of Gravity

ILLUSTRATION: Draw and label diagrams of a refracting and reflecting telescope in the appropriate boxes. *8 points*

REFRACTING TELESCOPE	REFLECTING TELESCOPE

_____ _____ ____/____/____
Student's Signature Parent's Signature Date

EARTH, MOON, SUN AND SEASONS

Teacher's Classwork Agenda and Content Notes

Classwork Agenda for the Week

1. Students will demonstrate that the earth rotates on its axis.

2. Students will explain the cause of earth's seasons.

3. Students will explain the cause of lunar and solar eclipses and calculate the diameter of the sun.

4. Students will show why the moon has phases.

Content Notes for Lecture and Discussion

The Copernican Revolution that began with the publication of **Nicolaus Copernicus's** (b. 1473; d. 1543) *On the Revolution of the Heavenly Spheres* did not end until the publication of **Sir Isaac Newton's** (b. 1642; d. 1727) *Principia* in 1687. In his *Principia*, Newton made the final synthesis of the physical and astronomical disciplines by showing the consistency between the laws of **Johannes Kepler** (b. 1571; d. 1630) and the enlightening telescope observations of **Galileo Galilei** (b. 1564; d. 1642). Until Newton put the debate to rest, Copernicus's idea was viewed as little more than a complex mathematical exercise with no basis in physical reality. Newton's definitive incorporation of the properties of mass, inertia, and momentum into the movement of the heavenly bodies made the Heliocentric Theory concrete. What followed was a deluge of new and numerous observations that laid the basis for modern astronomy.

With the invention of photography in the middle of the 19th century, concentration on the features and behavior of the sun and moon led to an increased understanding of the earth and its place in the cosmos. This, combined with Sir Isaac Newton's analysis of the nature of light, resulted in the study of **spectroscopy**. The English scientist **William Wollaston** (b. 1766; d. 1828) recorded the first **solar spectrum** in 1802. His record was interpreted in 1814 by the German physicist **Joseph von Fraunhofer** (b. 1787; d. 1826) who suggested that such spectra might give clues to the chemical composition of the sun. In 1859, the German physicists **Gustav Robert Kirchoff** (b. 1824; d. 1887) and **Robert Wilhelm Bunsen** (b. 1811; d. 1899) recognized the similarity between the solar spectrum and the spectra of burning gases in the laboratory. Together, they built the first **spectroscope** and used it to deduce the chemical composition of light emitted from a variety of sources including the sun and planets.

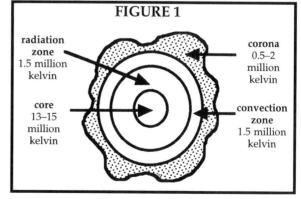

FIGURE 1

radiation zone
1.5 million kelvin

corona
0.5–2 million kelvin

core
13–15 million kelvin

convection zone
1.5 million kelvin

The **sun** is recognized today to be a medium-sized star of average brightness powered by the nuclear fusion of the element hydrogen into helium. The sun is approximately 1,395,000 kilometers in diameter compared to the 12,756 km diameter of the earth. It is tilted about 7° from the ecliptic (e.g., the path of earth's orbit) and rotates on its axis once in about 9 hours. Figure 1 is a simple diagram of the sun's basic architecture. The sun's surface **photosphere** and outlying **chromosphere** is teeming with violent particle and electromagnetic storms: **sunspots, eruptive prominences, polar plumes,** and **coronal streamers**. About 90% of the energy that flows to the surface is generated at the core which contains nearly 40% of the sun's total mass. Although the sun's volume is nearly

1,000,000 times that of earth, it is only about 330,000 times more massive. The sun is, therefore, less dense than earth which is the densest chunk of matter in the Solar System.

The most obvious surface features of the **moon** are its **craters**, **plains**, and **mountain ranges**. The moon has a diameter of 3,476 km and approximately one-sixth the gravitational attraction of earth at its surface. The moon has an iron core, a molten mantle, and silicate crust much like that of earth. It orbits the earth at an 8° incline with respect to the earth's ecliptic. The incline in the moon's orbit makes the monthly full moon possible when earth is nearer the sun. It is believed that geological action has virtually ceased on the moon since its formation 4.5 billion years ago; although seismometers placed there by Apollo astronauts in the 1970s have registered quakes that are probably the result of gravita-

tional forces exerted at its surface. The moon rotates once on its axis in the same period it takes to revolve around the earth, so the same face of the moon always looks to earth. A **sidereal month** is the time it takes for the moon to make a 360° orbit around the earth until both earth and moon are aligned with the same distant stars (e.g., 27.3 days). A **synodic month** is the time it takes for the moon to realign itself with the earth and

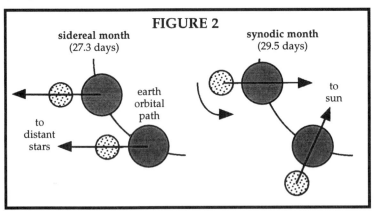

FIGURE 2

sidereal month (27.3 days)

synodic month (29.5 days)

earth orbital path

to distant stars

to sun

sun as it follows our planet in its journey around our star. Figure 2 illustrates the difference between a sidereal and synodic month.

The tilt of the earth in its revolution around the sun determines the change of seasons. The changing positions of the earth, moon, and sun with respect to one another results in lunar and solar eclipses as well as the changing phases of the moon.

In Lesson #1, students will demonstrate that the earth rotates on its axis.

In Lesson #2, students will explain the cause of earth's seasons.

In Lesson #3, students will explain the cause of lunar and solar eclipses and calculate the diameter of the sun.

In Lesson #4, students will show why the moon has phases.

ANSWERS TO THE HOMEWORK PROBLEMS

Students' illustrations should be reproductions of the illustrations they copied in class.

ANSWERS TO THE END-OF-THE-WEEK REVIEW QUIZ

1. true
2. rotates
3. pendulum
4. 27.3 or 29.5
5. were

6. true
7. 390,000
8. true
9. 365.25
10. astronomical

11. true
12. heliopause
13. medium-sized
14. helium
15. true

16. is
17. 11
18. solar wind
19. tilt
20. true

(A)	(B)	(C)	(D)	(E)
last quarter	waning gibbous	full	waxing gibbous	first quarter

EA11 FACT SHEET

EARTH, MOON, SUN AND SEASONS

CLASSWORK AGENDA FOR THE WEEK

(1) Demonstrate that the earth rotates on its axis.
(2) Explain the cause of earth's seasons.
(3) Explain the cause of lunar and solar eclipses and calculate the diameter of the sun.
(4) Show why the moon has phases.

Nicolaus Copernicus, a Polish-born scientist (b. 1473; d. 1543), changed the way people think about the universe. He proved that the earth **revolves** around the sun at a time when most people believed the earth was the center of the universe. Copernicus's idea was strange to most people because it challenged their direct observations. After all, we wake up every morning, see the sun rise and move across the sky, then dip over the horizon at dusk. At night, we watch the moon move across the sky hour after hour and see the stars travel in counterclockwise circles around the fixed North Star. It appears that the whole univese is revolving around us! In the previous unit—*Mapping the Heavens*—you used some of the same demonstrations Copernicus used to explain his theory. Using those and other demonstrations Copernicus suggested that the earth **rotates** on its **axis** once every 24 hours which causes day and night. But it was not until 1851 that the French physicist **Jean Bernard Léon Foucault** (b. 1819; d. 1868) proved beyond a doubt that the earth does rotate. He did this using a simple swinging pendulum. Copernicus explained that the **moon** was a **satellite** of earth and made one full trip around our planet about once every thirty days. He showed that the sun is the center of our solar system, shedding light on earth and on the other eight distant planets.

Ancient people reasoned that the earth was a sphere long before Columbus sailed to America. They watched ships "sink" beyond the horizon and return home, indicating the earth's surface was curved and not flat. They studied the shadow of the earth that was cast on the moon during lunar eclipses, noting that the shadows were always round. The Greek geographer and mathematician **Eratosthenes** (b. 276 B.C.; d. 194 B.C.) calculated the size of the earth in the third century B.C. and was accurate to within a few hundred miles.

Later astronomers like **Galileo Galilei** (b. 1564; d. 1642) made even more accurate calculations to determine the sizes of the moon, sun, and planets. Galileo discovered that the moon was not smooth but was covered with craters and rugged mountains. Using parallax, astronomers found the average distance to the moon to be **242,000 miles** (or about **390,000 kilometers**). They were able to explain **lunar** and **solar eclipses** as well as the **phases of the moon** (e.g., **crescent, quarter,** and **gibbous**) after studying the changing position of the moon relative to the earth and sun. Galileo discovered sunspots moving across the surface of the sun, suggesting that the sun—like earth—rotates on an axis. The earth is about **93,000,000 miles** (or **150,000,000 kilometers**) from the sun and completes one revolution around the sun every **365.25 days** (e.g., 1 year). The average distance from the earth to the sun is called one **astronomical unit** (or **a.u.**). The dimensions of the Solar System can be expressed in astronomical units. The most distant planet, Pluto, orbits at about 39.4 a.u. from the sun. The outer limit of the Solar System at about 100 a.u. from the sun is called the **heliopause**.

The sun is a medium-sized star of average brightness. The sun's **radiation** (e.g., light, heat, and ultraviolet rays) is produced in its dense, hot **core**. In the sun's core **nuclear fusion** combines **hydrogen** atoms to form **helium** atoms at the rate of hundreds of million metric tons per second. Even at this incredible rate of consumption the sun has enough hydrogen to burn for another 6 billion years. As dependable as it is, however, the sun is very active. Dark **sunspots** accompanied by violent **magnetic**

storms appear on the sun's surface in 11 year cycles. These storms give rise to gigantic bursts of energy that cause energetically charged atomic particles to flow out into space. This flow of energetic particles can effect the earth's weather, and radio and television communications. It can knock satellites out of their orbit and injure astronauts working in space. This blizzard of dust, gas, and energy is called the **solar wind**. The solar wind supplies our planet with enough energy to keep all of us alive. It drives the weather and can alter the climate. Today we know that the change of seasons is caused by the shift in the tilt of the earth as it revolves around the sun on its voyage through the storm of solar wind.

Homework Directions

1. Draw and label the positions of the earth, sun, and moon during a lunar and solar eclipse. Label the umbra and penumbra in each diagram.

2. Draw and label the positions of the earth and sun at summer solstice, autumnal equinox, winter solstice, and vernal equinox. Show the correct tilt of the earth in each position.

Assignment due: _____

_____   _____   ____/____/____
Student's Signature   Parent's Signature   Date

EARTH, MOON, SUN AND SEASONS

Work Date: ____/____/____

LESSON OBJECTIVE

Students will demonstrate that the earth rotates on its axis.

Classroom Activities

On Your Mark!

Begin discussion by asking students the following question: How do you know the earth rotates? They will give a variety of explanations that boil down to the fact that "someone told them it does." In the 15th century, people believed that the earth was fixed in space and that the universe revolved around us! Have students refer to paragraph #1 on their Fact Sheet to see why ancient people believed that the earth was the center of the universe. As good parents, they taught that belief to their children because it made sense and was based on their everyday observations. Yet today, we know that the earth is not the center of things; so, being told what to believe is not sufficient proof. How can we prove the earth rotates?

Get Set!

Introduce the French physicist **Jean Bernard Léon Foucault** (b. 1819; d. 1868) who proved with the use of a simple pendulum that the earth rotated on its axis. In 1851, Foucault suspended a heavy weight from a long wire into a large bowl at the Demonstration Hall in the Panthéon in Paris. He set the pendulum swinging so that it made contact with a series of pegs lined up around the circumference of the bowl. Draw Illustration A on the board to show the results of Foucault's demonstration. Point out that the pegs along the circumference of the bowl—which was fixed to the floor—were knocked over one-by-one by the swinging pendulum. The pendulum appeared to shift its trajectory. Foucault reasoned that the momentum of the pendulum must keep it moving in the same direction, back and forth according to Newton's First Law of Motion: "A body in motion will remain in motion unless acted upon by an outside force." So, it was the

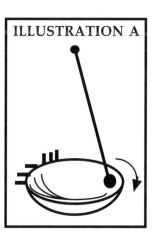

ILLUSTRATION A

earth beneath the swinging pendulum that was rotating to change the position of the pegs.

Go!

Give students ample time to perform activities described in Figure A on Journal Sheet #1. When they rotate the construction paper in the first activity the momentum of the weight will change slightly, altering the direction of swing. Point out that Foucault's pendulum was suspended by a very long string to eliminate this extraneous movement. Stress the "stubborness" of the pendulum and its resistance to change direction. Emphasize the fact that the rotation of the underlying surface does little to change the swing of the pendulum. Remind students of Foucault's results at the Panthéon in Paris.

Use the information in the Teacher's Classwork Agenda and Content Notes to give students a list of basic facts about the earth and sun. Have them list this information on Journal Sheet #1.

Materials

12" rulers, tape, 1-hole rubber stoppers, construction paper, string, 100 gram weights

Name: _____ Period:_____ Date: ____/____/____

EA11 JOURNAL SHEET #1

EARTH, MOON, SUN AND SEASONS

FIGURE A

<u>Directions:</u> (1) Make a "teepee frame" out of four 12" rulers, tape, and a 1-hole rubber stopper as shown. (2) Tie a 100 gram weight from a string and suspend it from the hole of the stopper. (3) Tape the frame to a piece of construction paper. (4) Set the weight swinging like a pendulum so that it swings from one ruler to the opposite ruler. (5) Gently rotate the teepee by moving the construction paper under it. (6) Record your observations. How does the pendulum behave? Does the swing remain between the two rulers as you rotate the construction paper or does it change the position of its swing?

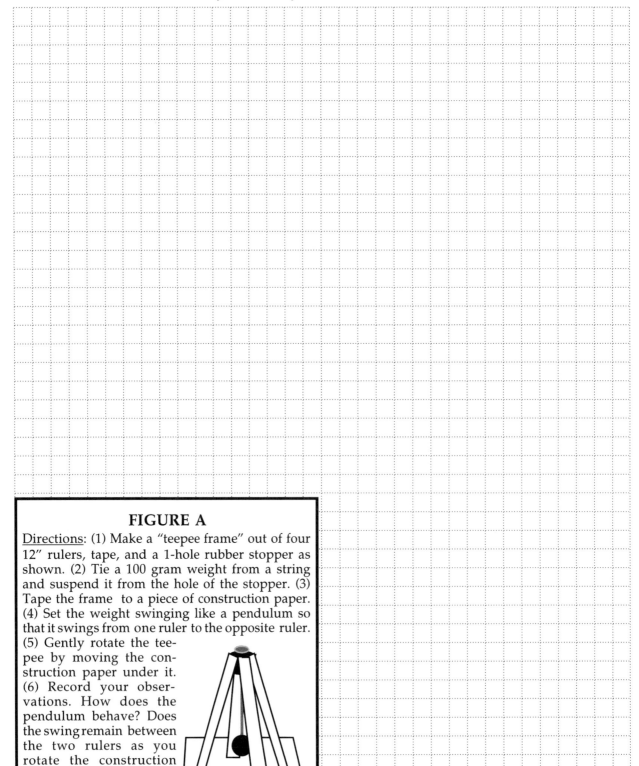

EARTH, MOON, SUN AND SEASONS

Work Date: ____/____/____

LESSON OBJECTIVE

Students will explain the cause of earth's seasons.

Classroom Activities

On Your Mark!

Begin discussion with a review of the causes of the weather and changes in earth's climate introduced in Units #7, #8, and #9. Remind students that the sun is the primary engine that drives changes in our atmosphere by heating the varied surfaces of the earth differentially. Draw Illustration A on the board and have students copy your drawing on Journal Sheet #2. Define the term **ecliptic** as a "horizontal plane, parallel to the imaginary floor of the universe, around which the earth orbits the sun." Note that our planet's axis is tilted 23.3° from the perpendicular to the ecliptic. Point out that the earth "wobbles" slightly on its axis like a spinning top that is slowing down. But for now, the North Pole is pointed toward the North Star. In the year 14,000 A.D., the earth's axis will point toward the star Vega in the constellation Lyra (e.g., small harp).

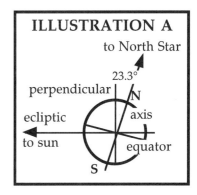

ILLUSTRATION A

to North Star

23.3°

perpendicular

ecliptic

to sun

N

axis

equator

S

Get Set!

Draw Illustration C and have students copy your drawing on Journal Sheet #2. Point out that the sun's direct rays hit different parts of the earth at different times of the year. The earth remains tilted toward the North Star throughout its revolution of the sun. At **winter solstice** in the Northern Hemisphere, the most direct rays of the sun hit the **tropic of capricorn** in the Southern Hemisphere. At **summer solstice** in the Northern Hemisphere, the most direct rays of the sun hit the tropic of cancer in the Northern Hemisphere.

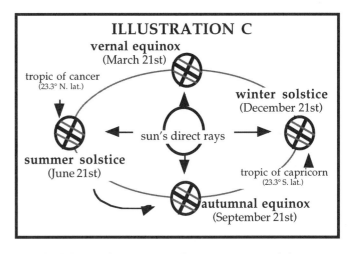

ILLUSTRATION C

vernal equinox
(March 21st)

tropic of cancer
(23.3° N. lat.)

winter solstice
(December 21st)

sun's direct rays

summer solstice
(June 21st)

tropic of capricorn
(23.3° S. lat.)

autumnal equinox
(September 21st)

At the **vernal** and **autumnal equinoxes** both hemispheres receive the same amount of direct sunlight. Seasons in the Northern and Southern hemispheres are, therefore, reversed.

Go!

Give students ample time to perform the activity described in Figure B on Journal Sheet #2. The student's graphs of temperature readings will show that the thermometer in direct line with the heat of the desk lamp warms more quickly.

Materials

desk lamp, ringstand and clamps, tape, thermometers

EA11 JOURNAL SHEET #2

EARTH, MOON, SUN AND SEASONS

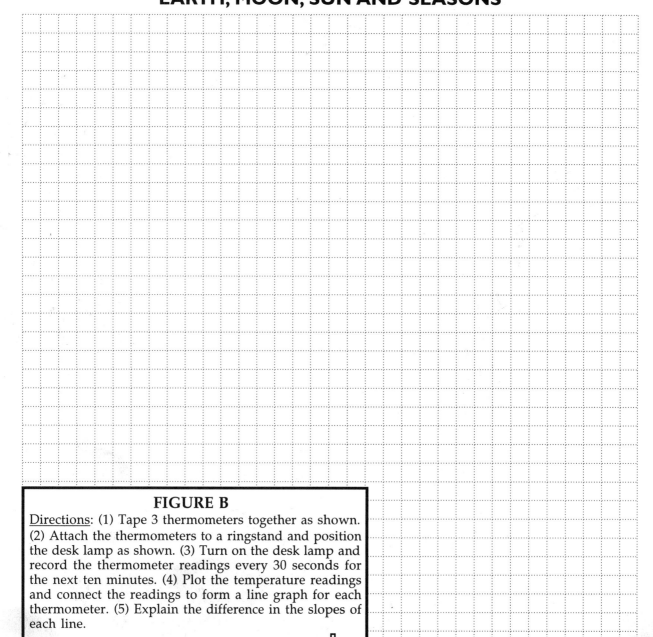

FIGURE B

<u>Directions</u>: (1) Tape 3 thermometers together as shown. (2) Attach the thermometers to a ringstand and position the desk lamp as shown. (3) Turn on the desk lamp and record the thermometer readings every 30 seconds for the next ten minutes. (4) Plot the temperature readings and connect the readings to form a line graph for each thermometer. (5) Explain the difference in the slopes of each line.

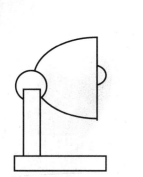
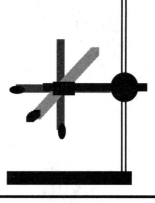

EA11 Lesson #3

EARTH, MOON, SUN AND SEASONS

Work Date: ____/____/____

LESSON OBJECTIVE

Students will explain the cause of lunar and solar eclipses and calculate the diameter of the sun.

Classroom Activities

On Your Mark!

Draw Illustration D on the board and ask the following question: "If these objects were exactly the same size and Object A were the sun, what kind of shadow would 'B' cast on 'C' "? Answer: Object C would be completely covered with the shadow of B and vice versa if positions were reversed. Point out that shadows cast on the earth by the moon during a solar eclipse never completely cover the earth because the earth is larger than the moon yet very much smaller than the sun. Draw Illustration E on the board to show how the Greek astronomer **Eratosthenes** (b. 276 B.C.; d. 194 B.C.) calculated the size of the earth. He examined the shadows cast by vertical sticks put in the ground in Egypt nearer the equator and in Greece farther north. He noted that the rays of the sun could be compared to two parallel rays drawn through a circle. He also noted that both sticks placed perpendicular with the earth's surface must point directly toward the center of the earth. The laws of geometry dictate that "angle a" is equal to "angle b." He measured angle b (e.g., the angle of the shadow cast from the top of the stick in Greece). It was one-fiftieth that of a full circle. So, angle a would also have to be one-fiftieth of a circle. Eratosthenes reasoned that the circumference of the earth was therefore fifty times the distance from the stick in Egypt to the stick in Greece which he knew to be about 500 miles: $50 \times 500 = 25,000$ miles. The actual circumference of the earth is about 24,833 miles.

ILLUSTRATION D

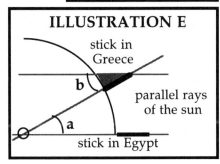
ILLUSTRATION E

stick in Greece

parallel rays of the sun

b

a

stick in Egypt

Get Set!

Draw Illustration F and have students copy your drawing on Journal Sheet #3. Point out that the sun's direct rays are obscured during **lunar** and **solar eclipses** resulting in shadows being cast on the moon by the earth and the earth by the moon, respectively. The dark shadow called an **umbra** is the area where no light from the sun reaches the surface. The **penumbra** is the area surrounding the umbra where some light reaches the surface. Ask students to consider their own shadow. Do they sometimes cast a dark shadow surrounded by a lighter one?

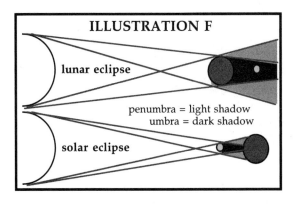
ILLUSTRATION F

lunar eclipse

penumbra = light shadow
umbra = dark shadow

solar eclipse

Go!

Give students ample time to perform the activity described in Figure C on Journal Sheet #3. They should arrive at an answer of about 1.4×10^{12} millimeters.

Materials

metric rulers, index cards, insect pins, the sun

EA11 JOURNAL SHEET #3

EARTH, MOON, SUN AND SEASONS

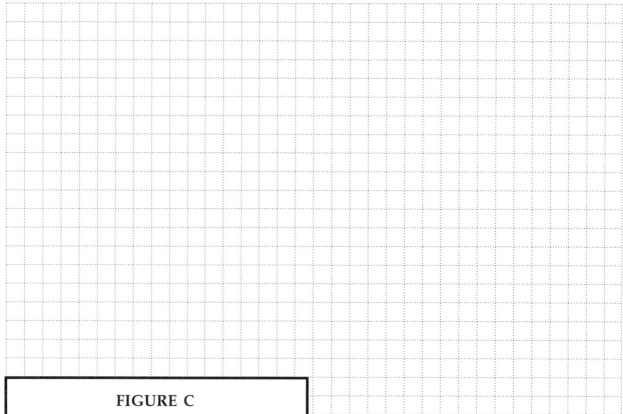

FIGURE C

Directions: (1) Tape an index card to one end of a meter stick. (2) Cut another index card so that it can slide along the length of the meter stick as shown. (3) Use an insect pin to punch a small hole in the center of the "sliding" index card. (4) Place the meter stick over your shoulder and face away from the sun. (5) Align the two index cards with the sun and move the sliding card back and forth until a small patch of light from the sun appears on the fixed card. (6) Focus the image as best as you can. (7) Record the distance between the two cards to the nearest millimeter and circle the sun's image with a pencil exactly as it appears on the fixed card. (8) Calculate the diameter of the sun with the following formula:

$$D_s = \frac{D_i + S_s}{S_i}$$

where D_s is the diameter of the sun, D_i is the diameter of the sun's focused image on the card, S_s is 150,000,000,000,000 millimeters, and S_i is the distance from the sliding card to the focused image.

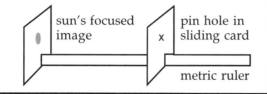

sun's focused image   pin hole in sliding card

metric ruler

EARTH, MOON, SUN AND SEASONS

Work Date: ____/____/____

LESSON OBJECTIVE

Students will show why the moon has phases.

Classroom Activities

On Your Mark!

Draw Illustration G on the board and have students copy your drawing on Journal Sheet #4. Explain that the moon's orbit is "slanted" with respect to the ecliptic. Remind them, also, that the sun—having a diameter of nearly 1.5 million kilometers—is very much larger than the earth. These two factors allow the moon to receive light from the sun even when the earth is closer to the sun than the moon. The moon can be "behind" the earth but also "above or below" it. This allows us to see a full moon once a month. Of course, the revolution of the moon around the earth—as the earth revolves around the sun—also produces eclipses from time to time.

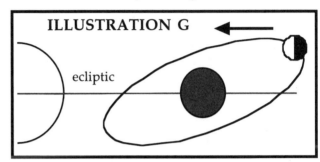

ILLUSTRATION G

ecliptic

Get Set!

Draw Illustration H on the board and have students copy your drawing onto Journal Sheet #4. Point out that this illustration represents a view from the earth's North Pole. It is a "polar projection." From this viewpoint the moon orbits in a counterclockwise direction (e.g., from new moon to waxing crescent, etc.). Students need to put themselves in the position of the earth as they look at the moon "overhead." Light is "waxed on" to the moon as it makes its journey to a full moon at midmonth. Light is "wiped off" or "waning" as the moon completes its orbital trip around the earth.

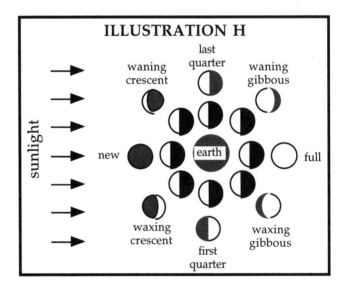

ILLUSTRATION H

last quarter

waning crescent

waning gibbous

sunlight

new

earth

full

waxing crescent

waxing gibbous

first quarter

Go!

Give students ample time to perform the activity described in Figure D on Journal Sheet #4.

Materials

flashlights, large balloons or balls, tape

EA11 JOURNAL SHEET #4

EARTH, MOON, SUN AND SEASONS

FIGURE D

Directions: (1) Use tape to secure an inflated balloon (or large ball) to the center of a table. (2) Darken the room and shine a flashlight on the balloon so that the entire side facing you is lit. (3) Have a classmate stand behind you so that you can both observe the completely illuminated face of the balloon. (4) Have your classmate walk around the table and report the amount and shape of the patch of light illuminating the balloon. Do the patches of light and shadow on the balloon change shape? Are the patches curved, crescent, or straight and vertical? Compare your observations to the pictures you drew of the changing phases of the moon.

EA11 Review Quiz

Directions: Keep your eyes on your own work.
Read all directions and questions carefully.
THINK BEFORE YOU ANSWER!
Watch your spelling, be neat, and do the best you can.

CLASSWORK	(~40):	_____
HOMEWORK	(~20):	_____
CURRENT EVENT	(~10):	_____
TEST	(~30):	_____
TOTAL	(~100):	_____

(A ≥ 90, B ≥ 80, C ≥ 70, D ≥ 60, F < 60)

LETTER GRADE: _____

TEACHER'S COMMENTS: _____

_____

EARTH, MOON, SUN AND SEASONS

TRUE–FALSE FILL-IN: If the statement is true, write the word TRUE. If the statement is false, change the underlined word to make the statement true. *20 points*

_____ 1. Nicolaus Copernicus proved that the <u>sun</u> is the center of the universe.

_____ 2. Copernicus suggested that the earth <u>revolves</u> on its axis once every 24 hours which causes day and night.

_____ 3. The French physicist Foucault used a(n) <u>astrolabe</u> to prove that the earth rotates on its axis.

_____ 4. The moon makes one full trip around our planet about once every <u>365.25</u> days.

_____ 5. Ancient Greeks <u>were not</u> aware that the earth was a sphere.

_____ 6. The Greek geographer and mathematician <u>Eratosthenes</u> calculated the size of the earth in the third century B.C.

_____ 7. Astronomers have found the average distance to the moon to be <u>242,000</u> kilometers.

_____ 8. The earth is about 150,000,000 <u>kilometers</u> from the sun.

_____ 9. The earth completes one revolution around the sun every <u>30</u> days.

_____ 10. The average distance from the earth to the sun is called one <u>solar</u> unit.

_____ 11. The most distant planet, Pluto, orbits at about <u>39.4</u> a.u. from the sun.

_____ 12. The outer limits of the Solar System at about 100 a.u. from the sun is called the <u>end of the universe</u>.

_____13. The sun is a(n) <u>very large</u> star of average brightness.

_____14. In the sun's core, hydrogen atoms fuse to form <u>oxygen</u> atoms.

_____15. The sun has enough hydrogen to burn for another 6 <u>billion</u> years.

_____16. The sun <u>is not</u> very active.

_____17. Dark sunspots accompanied by violent magnetic storms appear on the sun's surface in <u>1,000</u> year cycles.

_____18. The blizzard of dust, gas, and energy streaming from the sun toward earth is called the <u>heliopause</u>.

_____19. The change of seasons is caused by the shift in the <u>speed</u> of the earth as it revolves around the sun.

_____20. The shadow of the earth is cast on the moon during a <u>lunar</u> eclipse.

ILLUSTRATION: In each empty dotted circle, draw and label the moon phase as it appears from earth in each of the shown positions. Assume that no eclipses will occur during this orbital period. *10 points*

Name the phase

A. _____

B. _____

C. _____

D. _____

E. _____

_____ _____ ___/___/___

THE HISTORY AND FUTURE OF SPACE TRAVEL

TEACHER'S CLASSWORK AGENDA AND CONTENT NOTES

Classwork Agenda for the Week

1. Students will demonstrate the principle of rocketry and the gyroscope.
2. Students will review the accomplishments of manned space missions.
3. Students will discuss the advantages and disadvantages of unmanned space probes.
4. Students will debate issues relevant to the future of the space program.

Content Notes for Lecture and Discussion

The beginning of space exploration was impeded by the technological difficulties associated with the development of vehicles that had the sheer power to escape earth's gravitational pull and surpass the **escape velocity** of the planet (e.g., orbital velocity = 27,360 km per hour; escape velocity = 40,323 km per hour). The Russian scientist **Konstantin Eduarovich Tsiolkovsky** (b. 1857; d. 1935) was among the first to calculate the escape velocity of earth and develop the first coherent theory of spaceflight. In 1903, Tsiolkovsky suggested the use of liquid propellants such as liquified oxygen to overcome the obvious problems of heavy solid fuels, like gunpowder, previously used in military rocket applications. He also introduced the idea of the **multistage rocket**: a vehicle designed to lose weight by discarding spent fuel stages thereby increasing the ascending vehicle's velocity. Following the work of American scientist **Robert Hutchings Goddard** (b. 1882; d. 1945), German aerospace engineer **Werner von Braun** (b. 1912; d. 1977) developed the **V-2 Rocket** for use against England during World War II. After the war, von Braun assisted American scientists in the development of military and civilian space vehicles.

The space race between the United States and the Soviet Union that began in the late 1950s with the launch of **Sputnik 1** in 1957 virtually ended with the American manned-landing on the moon by **Apollo 11** in 1969. Following the successful landings on the moon by the Americans, the Soviet Union confined itself to the development of their **Soyuz**, **Salyut**, and **Mir space stations**. To date, Russian cosmonauts hold the records for long-term work in orbit. The Americans, on the other hand, are unsurpassed in their manned exploration of the moon and unmanned probing of the planets and deep space. The number of missions into space number in the hundreds, and the number of **communication**, **weather**, and **spy satellites** comes close to a thousand. In fact, scientists worry about the mounting space debris that encircles the planet (e.g., loose nuts and bolts, wires, and small solar panels left in orbit by working astronauts) spurring fears of a mishap that it could cause. The problem has produced a number of serious studies on the subject.

Manned missions, of course, have the advantage of putting "quick thinking people" where the problems arise. People will be needed in space until robots can be designed to solve problems "on the spot" without the delay associated with radio command-and-control at a distance (e.g., at the speed of light taking hours from one end of the solar system to the other). Of course, the cost of sending humans into space is exorbitant because of the precautions needed to protect life and limb. **Space probes**, therefore, are the safer and more economical choice when it comes to the acquisition of formative data. To date, space probes have gathered essential information about the sun, moon, and planets in our solar system and future missions promise to increase that store of knowledge.

EA12 Content Notes *(cont'd)*

The exploration of space is consistent with humankind's insatiable desire for knowledge and will serve to satisfy our curiosity only when we have "been there." Like the explorers of the 15th and 16th centuries, future astronauts will search out the mysteries of the celestial sphere just as Columbus and Magellan sought to resolve their picture of the globe.

In Lesson #1, students will demonstrate the principles of rocketry and the gyroscope.

In Lesson #2, students will research and review the accomplishments of manned space missions and report their findings to their classmates.

In Lesson #3, students will research and review the data gathered by unmanned space probes and discuss with their classmates their advantages and disadvantages over manned missions.

In Lesson #4, students will debate issues relevant to the future of the space program and write a brief summary of the debate stating their personal points of view.

ANSWERS TO THE HOMEWORK PROBLEMS

Stories will vary but should show the students' appreciation of the unique environment of space. Tasks performed inside a space station are complicated primarily by the absence of gravity (e.g., moving about, sleeping, bathing, eating, etc.) Space walks involve the added complications of a space suit which is—for all intents and purposes—a strictly self-contained environment that must protect the astronaut against the vacuum and intense radiation of space.

ANSWERS TO THE END-OF-THE-WEEK REVIEW QUIZ

1. Montgolfier	6. rockets	11. C
2. Wright	7. Sir Isaac Newton	12. B
3. true	8. R.H. Goddard	13. D
4. true	9. gyroscope	14. E
5. true	10. W. von Braun	15. A

ESSAY #1: For every action there is an equal and opposite reaction.

ESSAY #2: Answers will vary but should demonstrate the student's grasp of some of the issues discussed in class.

EA12 FACT SHEET

THE HISTORY AND FUTURE OF SPACE TRAVEL

CLASSWORK AGENDA FOR THE WEEK

(1) Demonstrate the principle of rocketry and the gyroscope.
(2) Review the accomplishments of manned space missions.
(3) Discuss the advantages and disadvantages of unmanned space probes.
(4) Debate issues relevant to the future of the space program.

Humans have always been impressed by the ability of birds to fly. The first successful flight of a hot air balloon in 1783 by French brother-inventors **Joseph** (b. 1740; d. 1810) and **Jacques Montgolfier** (b. 1745; d. 1799) strengthened people's desire to explore the outer reaches of the atmosphere. In 1903, American brother-inventors **Orville** (b. 1871; d. 1948) and **Wilbur Wright** (b. 1867; d. 1912) made the first successfully controlled flight of a gasoline powered airplane. Since then, engineers and pilots have designed faster and faster aircraft because a trip into outer space requires an enormous amount of speed.

Scientists have known for some time that a trip into orbit requires a speed of 27,360 kilometers per hour (e.g., 16,960 miles per hour). To escape earth's gravity and make a journey to the moon or beyond requires speeds of greater than 40,323 kilometers per hour (e.g., 25,000 miles per hour). The speed needed to escape earth's gravity is called the **escape velocity** of earth. The science fiction writer **Jules Verne** (b. 1828; d. 1905) imagined explorers launching themselves into space using an immense cannon. Cannon balls were the fastest moving objects in the author's time. The Chinese invented rockets as early as 1,000 A.D. which they used in their fireworks displays on holidays and other special occasions. They used gunpowder to fuel their rockets. But rocketry didn't become a real science until the 18th century.

The principal of rocketry was first explained by **Sir Isaac Newton** (b. 1642; d. 1727) in his **Third Law of Motion**. The law can be stated as follows: *For every action there is an equal and opposite reaction.* The gases produced by the burning fuel of a rocket are expelled with considerable force. The force of the escaping gas exerts an "equal and opposite" force that pushes the rocket into the sky. In 1926, the American rocket pioneer **Robert Hutchings Goddard** (b. 1882; d. 1945) launched the first liquid-fueled rocket; and by 1935, he had equipped his rockets with **gyroscopes** to keep them in balance and recording instruments to make measurements of the upper atmosphere. After World War II, the German scientist **Werner von Braun** (b. 1912; d. 1977) helped American scientists to build a rocket capable of launching a **payload** into orbit. By the start of the 1960s, both the Americans and Russians were in a feverish race into space.

The first human launched into orbit was the Russian **cosmonaut Yuri Gagarin** (b. 1938; d. 1968) who made his historic flight in 1961. The first American in space was astronaut **Alan Bartlett Shepard** (b. 1923; d. 1998). Shepard was also the fifth person to take a stroll on the surface of the moon. In 1963, the Russians launched the first woman into space: **Valentina Vladimirovna Tereshkova** (b. 1937). The first person to walk on the moon was the American astronaut **Neil Armstrong** (b. 1930). After Armstrong, the surface of the moon was revisited six times between 1969 and the end of 1972. The first American woman to ride the **Space Shuttle** was **Sally Kirsten Ride** (b. 1951).

EA12 Fact Sheet (cont'd)

The cost of manned space exploration is extremely expensive. So, unmanned **probes** are used to do the basic work of gathering information about earth's planetary neighbors. Since the early 1960s, scores of space probes have been sent to explore the planets. And as recently as the 1980s, two have left our solar system after probing the atmospheres of Jupiter, Uranus, Neptune, and Saturn: **Voyager I** and **Voyager II** launched in 1977 and 1980. At present, scientists and engineers are working around the clock to plan and build a **space station** that will house dozens of astronauts in space. And, the **National Aeronautics and Space Administration** (NASA)—the government organization that schedules and carries out space missions—is planning a manned trip to Mars by the second decade of the 21st century.

Homework Directions

Write a brief short story of no more than 250 words describing what it might be like to work on a space station. Think of a simple activity you might perform at home and describe the ease or difficulty you might have completing the same activity in the weightlessness of space.

Assignment due: _____

_____   _____   ____/____/____
Student's Signature   Parent's Signature   Date

THE HISTORY AND FUTURE OF SPACE TRAVEL

Work Date: ____/____/____

LESSON OBJECTIVE

Students will demonstrate the principle of rocketry and the gyroscope.

Classroom Activities

On Your Mark!

Use the Teacher's Classwork Agenda and Content Notes and the students' Fact Sheet to point out the primary obstacle to space travel: **the escape velocity of earth**. Point out that rockets—invented by the Chinese as early as 1,000 A.D.—were considered the vehicles most likely able to achieve the high rates of speed necessary to put a craft into orbit. Review the work of **Tsiolkovsky** (b. 1857; d. 1935), **Goddard** (b. 1882; d. 1945), and **von Braun** (b. 1912; d. 1977) in developing rocket technology in the first half of the 20th century. Explain the physical law that serves as the basic principle of rocketry: **Sir Isaac Newton's** (b. 1642; d. 1727) **Third Law of Motion**. Have students write that law on Journal Sheet #1: <u>For every action there is an equal and opposite reaction.</u>

Get Set!

Have students copy your drawing of Illustration A. Point out that—unlike jet aircraft that use the oxygen in the atmosphere to burn their propellant—rockets are self-contained vehicles that must carry both the **fuel** and **oxidizer** needed to burn their fuel. Explain that the **gyroscope** helps to keep the rocket from toppling off course. The first gyroscope was invented by engineers of Great Britain's Royal Navy in 1744. The principle of the gyroscope is based on **Newton's First Law of Motion** which asserts that an <u>object in motion will remain in motion unless acted upon by an outside force</u>. When a gyroscope is spinning its momentum resists any change in direction. So, gyroscopes can be used to "sense" when a rocket begins to topple one way or another. Engine direction and fins attached to the rocket's tail can be adjusted to correct the angle of tilt.

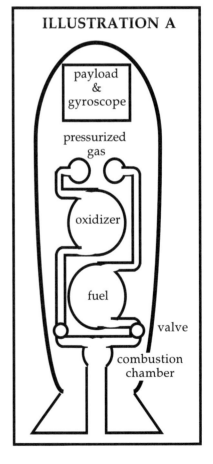

ILLUSTRATION A

payload & gyroscope

pressurized gas

oxidizer

fuel

valve

combustion chamber

Go!

Give students ample time to perform the activities described in Figure A and Figure B on Journal Sheet #1.

Materials

long balloons, straws, string, tape, tables and chairs, cardboard, stubby pencils, compass

EA12 JOURNAL SHEET #1

THE HISTORY AND FUTURE OF SPACE TRAVEL

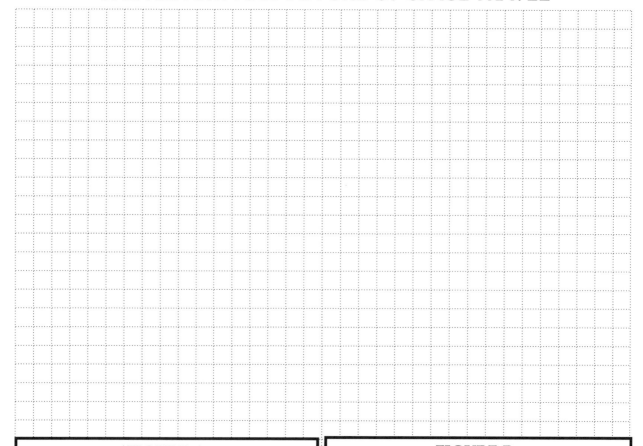

FIGURE A

<u>Directions</u>: (1) Anchor one end of a 5-meter length of string to a desk, table, or chair. (2) Thread the other end through a plastic straw and anchor that end 5 meters away so that the string is taut. (3) Tape the straw to a long balloon. (4) Stand at one anchor, inflate the balloon, and let it go. (5) Conduct a race with your classmates and discuss the factors that cause your rocket to accelerate. Discuss the factors that cause the rocket to slow down.

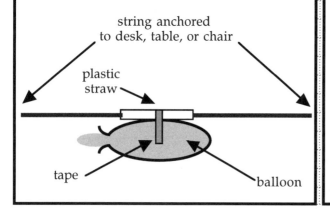

string anchored to desk, table, or chair

plastic straw

tape

balloon

FIGURE B

<u>Directions</u>: (1) Use a compass to draw a circle with a 15-centimeter diameter on a piece of cardboard. (2) Cut out the circle. (3) Poke a hole in the exact center of the circle with a stubby pencil and secure the pencil to the cardboard with tape. (4) Tie a slip knot in a string and loop it over the end of the pencil as shown. (5) Tape the string to the end of a table. (6) Hold both ends of the pencil between your thumb, fore, and middle fingers and give the disc a spin. (7) Record your observations. Does the disc remain vertical as it spins? What happens when it stops spinning? (8) Lift the disc back to a vertical position and spin it again. (9) Poke it as it spins with a pencil eraser. Does it resist your prodding? Explain your observations.

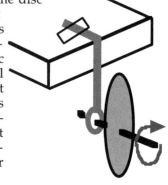

THE HISTORY AND FUTURE OF SPACE TRAVEL

Work Date: _____/_____/_____

LESSON OBJECTIVE

Students will review the accomplishments of manned space missions.

Classroom Activities

On Your Mark!

Prepare for class by checking out from the school library 10–20 reference books containing information about American- and Russian-manned spacecraft and missions. Give a brief overview of each of the <u>Manned Mission Projects</u> listed on Journal Sheet #2. American missions were primarily orchestrated around the goal set by **President John F. Kennedy** (b. 1917; d. 1963) in 1961 to put a man on the moon and bring him safely back to earth by the end of the 1960s. The Russians abandoned their efforts to go to the moon early in the race and concentrated on developing a permanent presence in space aboard their orbiting space stations.

Get Set!

Assign groups of cooperative students the task of compiling as much information as they can in the span of 20 minutes about a particular Manned Mission Project. Give them a large sheet of butcher or construction paper to use in creating a graphic organizer complete with quick sketches of the spacecraft involved in their particular project.

Go!

Give students a few minutes to quickly present their information to the class. Have attending student take notes on Journal Sheet #2.

Materials

butcher or construction paper, library reference materials, crayons, colored pencils

EA12 JOURNAL SHEET #2

THE HISTORY AND FUTURE OF SPACE TRAVEL

MANNED MISSION PROJECTS	
American	Russian
Mercury	Sputnik
Gemini	Vostok
Apollo	Soyuz
Skylab	Salyut
Space Shuttle	Mir

THE HISTORY AND FUTURE OF SPACE TRAVEL

Work Date: ____/____/____

LESSON OBJECTIVE

Students will discuss the advantages and disadvatanges of unmanned space probes.

Classroom Activities

On Your Mark!

Prepare for class by checking out from the school library 10–20 reference books containing information about the unmanned space probes listed on Journal Sheet #3. Give a brief summary of the advantages and disadvantages of <u>American Space Probes</u> as discussed in the Teacher's Classwork Agenda and Content Notes. Ask students to participate by adding or commenting on your presentation.

Get Set!

Assign groups of cooperative students the task of compiling as much information as they can in the span of 20 minutes about a particular space probe. Give them a large sheet of butcher or construction paper to use in creating a graphic organizer complete with quick sketches of the spacecraft and planet(s) it explored.

Go!

Give students a few minutes to quickly present their information to the class. Have attending student take notes on Journal Sheet #3.

Materials

butcher or construction paper, library reference materials, crayons, colored pencils

EA12 JOURNAL SHEET #3

THE HISTORY AND FUTURE OF SPACE TRAVEL

AMERICAN SPACE PROBES

Mariner
Viking
Pioneer
Voyager
Magellan
Galileo
Pathfinder
SOHO

THE HISTORY AND FUTURE OF SPACE TRAVEL

Work Date: _____/_____/_____

LESSON OBJECTIVE

Students will debate issues relevant to the future of the space program.

Classroom Activities

On Your Mark!

Explain to students that a debate begins with a **proposition**. The proposition states an opinion that the debaters may either support or oppose. Have students copy the following proposition (or any other relating to the topic of space travel that you feel would be of interest to them) on Journal Sheet #4: "Space exploration is expensive and the world has a lot of important problems that need to be addressed. The exploration of space should wait until all of these other problems are solved."

Get Set!

Divide the classroom in half and explain that one half of the class is in favor of the proposition while the other half is opposed to it. Assign these positions arbitrarily. Give students 10 minutes to consult with their groups, take notes on their position on Journal Sheet #4, then mix the class so that each new group contains two people in favor of the proposition and two people opposed to it.

Go!

Explain that a debate requires that each participant be given a set time to give evidence to support his or her own position in favor of or in opposition to the proposition. Set a timer to give each person two minutes to express a view before going on to the next person. Instruct students to take notes on the opposing point of view on Journal Sheet #4. At the end of 20 minutes students should return to their assigned seats and write a personal statement regarding their feelings about the proposition.

Materials

Journal Sheet #4, egg timer

EA12 JOURNAL SHEET #4

THE HISTORY AND FUTURE OF SPACE TRAVEL

THE PROPOSITION

EVIDENCE SUPPORTING THE PROPOSITION   EVIDENCE OPPOSING THE PROPOSITION

PERSONAL SUMMARY IN FAVOR OF OR IN OPPOSITION TO THE PROPOSITION

EA12 REVIEW QUIZ

Directions: Keep your eyes on your own work.
Read all directions and questions carefully.
THINK BEFORE YOU ANSWER!
Watch your spelling, be neat, and do the best you can.

TEACHER'S COMMENTS: _____
_____

THE HISTORY AND FUTURE OF SPACE TRAVEL

TRUE–FILL-IN: If the statement is true, write the word TRUE. If the statement is false, change the underlined word to make the statement true. *10 points*

_____ 1. The first successful flight of a hot air balloon was made by the <u>Wright</u> brothers.

_____ 2. In 1903, the first successfully controlled flight of a gasoline powered airplane was made by the <u>Montgolfier</u> brothers.

_____ 3. A trip into orbit requires a speed of about <u>27,000</u> kilometers per hour.

_____ 4. To escape earth's gravity and make a journey to the moon requires speeds of greater than <u>40,000</u> kilometers per hour.

_____ 5. The speed needed to escape earth's gravity is called the <u>escape velocity</u> of earth.

_____ 6. The Chinese invented <u>airplanes</u> as early as 1,000 A.D. which they fueled with explosive gunpowder

_____ 7. The principal of rocketry was first explained by <u>Albert Einstein</u> in his Third Law of Motion.

_____ 8. In 1926, the rocket pioneer <u>W. von Braun</u> launched the first liquid-fueled rocket.

_____ 9. A <u>periscope</u> helps to keep a rocket in balance.

_____ 10. After World War II, the scientist <u>R. H. Goddard</u> helped to build a rocket capable of launching a payload into orbit.

EA12 Review Quiz (cont'd)

MATCHING: Choose the letter of the accomplishment at right made by the person named at left. *10 points*

_____ 11. Gagarin   (A) first American woman in space

_____ 12. Tereshkova   (B) first woman in space

_____ 13. Shepard   (C) first man in space

_____ 14. Armstrong   (D) first American in space

_____ 15. Ride   (E) first man to walk on the moon

ESSAY #1: Write a sentence that summarizes the basic principle of rocketry. *5 points*

_____

_____

_____

ESSAY #2: Write several sentences that summarize the advantages of manned space missions and unmanned space probes. *5 points*

_____

_____

_____

_____

_____

_____

_____

_____

_____

THE INNER PLANETS

TEACHER'S CLASSWORK AGENDA AND CONTENT NOTES

Classwork Agenda for the Week

1. Students will compare and contrast Earth with the planet Mercury.
2. Students will compare and contrast Earth with the planet Venus.
3. Students will compare and contrast Earth with the planet Mars.
4. Students will compare and contrast the characteristics of comets, asteroids, and meteors.

Content Notes for Lecture and Discussion

Although the Danish astronomer **Tycho Brahe** (b. 1546; d. 1601) supplied astronomers with most of the observations they needed to substantiate the **heliocentric theory of the solar system**, he denied its truth until the day he died. So stubborn was Brahe—despite the overwhelming evidence that he himself had helped to collect—that he declared the Earth to be the only planet that did not revolve around the sun. The German astronomer **Johannes Kepler** (b. 1571; d. 1630) who worked as a student of Brahe later elucidated the patterns of elliptical motion exhibited by all the known planets. With the work of **Sir Isaac Newton** (b. 1642; d. 1727), the laws of motion and gravity governing the movement of the planets became clear. With the advance of the science of rocketry and the development of sufficient technology, modern scientists and engineers have explored the outer reaches of the solar system with a score of fantastic space probes. As a result of this work, much has been learned about the nature of our planetary neighbors and much, much more will be learned in years to come.

The facts presented in the chart below will assist you to make sure that students accumulate accurate information in their research about the planets. The "orbital eccentricity" of a planet's orbit refers to the planet's deviation from a perfect circle in its revolution around the sun. Eccentricity is calculated by the following formula: $e = 2a/2b$; where "e" is eccentricity, "a" is the distance between the two foci of the ellipse, and "b" is the length of the ellipse's long axis. As the

INNER PLANET BASIC FACTS				
planet	Mercury	Venus	Earth	Mars
avg. distance to sun (10^6 km)	57.9	108.2	149.6	227.9
orbital eccentricity	0.21	0.01	0.17	0.09
orbital inclination	7°	3.4°	0°	1.9°
period of revolution	88 days	224.7 days	365.25 days	687 days
period of rotation	59 days	243 days	23.9 hours	24.6 hours
orbital velocity (km/sec)	48	35	30	24
diameter at equator (km)	4,880	12,100	12,756	6,794
volume compared to earth	0.06	0.90	1.0	0.15
mass compared to earth	0.06	0.81	1.0	0.107
density compared to water	5.4	5.3	5.5	3.9
gravity compared to earth	0.38	0.91	1.0	0.38
escape velocity	4.2 km/s	10.5 km/s	11.3 km/s	5.2 km/s
albedo	0.06	0.8	0.36	0.15
known satellites	0	0	1	2

eccentricity of a planet's orbit approaches "0" the more circular is its orbit. As it approaches "1" the more elliptical the orbit is. The "orbital inclination" refers to the orbit of the planet in relation to earth's orbit (e.g., the ecliptic = 0°). Albedo is a measure of a planet's reflective properties. A planet's reflectance is calculated using the following formula: $a = c/d$; where "a" is albedo, "c" is the light reflected by the surface of the planet, and "d" is the amount of sunlight that falls on the planet. An albedo approaching "0" means the planet is highly "absorbent." An albedo approaching "1" means it is highly reflective.

In Lesson #1, students will compare and contrast Earth with the planet Mercury.

In Lesson #2, students will compare and contrast Earth with the planet Venus.

In Lesson #3, students will compare and contrast Earth with the planet Mars.

In Lesson #4, students will compare and contrast the characteristics of comets, asteroids, and meteors.

ANSWERS TO THE HOMEWORK PROBLEMS

Students should be prepared to present their string of the inner planets by the end of this unit. As 1 millimeter is equal to 100,000 kilometers, on the homework scale 1 centimeter equals 1,000,000 kilometers. The index card for the planet Mercury should be placed 58 cm from the sun; Venus should be placed 108 cm from the sun; the Earth, 150 cm; Mars 227 cm; and the asteroid belt between 300 to 450 cm from the sun.

ANSWERS TO THE END-OF-THE-WEEK REVIEW QUIZ

1. C
2. B
3. D
4. A
5. B

6. D
7. E
8. B
9. A
10. D

EA13 FACT SHEET

THE INNER PLANETS

CLASSWORK AGENDA FOR THE WEEK

(1)   Compare and contrast the Earth with the planet Mercury.
(2)   Compare and contrast the Earth with the planet Venus.
(3)   Compare and contrast the Earth with the planet Mars.
(4)   Compare and contrast the characteristics of comets, asteroids, and meteors.

The largest objects in the solar system are the sun, the nine planets orbiting the sun, and the moons that orbit most of the nine planets. The planets Mercury, Venus, Earth, and Mars comprise **the inner planets** of the system. Beyond Mars is the asteroid belt which contains thousands of large and small rocks ranging in size from several meters to several hundred kilometers across. All of the planets, and many of their moons, are named after the gods worshipped by ancient Greeks and Romans.

The planet **Mercury** is the closest planet to the sun, named for the "messenger" of the gods. Mercury is a difficult planet to study using a telescope because of its proximity to the glaring sun. It goes through phases like Earth's moon and the only time it is visible from Earth is moments before dusk. At dusk, Mercury becomes blurred by our own thick atmosphere. The Mariner 10 spacecraft sent back the first detailed photographs of Mercury's surface in 1974 and discovered the planet's intense magnetic field. Mercury is a barren world with hardly any atmosphere and is covered with craters like that of Earth's moon.

The planet **Venus** is the second closest planet to the sun, named for the goddess of "love and beauty." While Venus appears bright in the night sky almost every day of the year, her surface is completely obscured by dense swirling clouds. Mariner 10 sent back the first revealing photographs of Venus' violent atmospheric conditions before continuing on its journey to Mercury. It is nearly the same size as Earth and is sometimes called our "sister planet." Venus takes more time to rotate once on its axis than it does to make one complete revolution around the sun. This makes Venus the only planet in the solar system whose day is longer than its year! Early science fiction writers suspected that Venus was lush with vegetation and animal life like the tropical rain forests of earth. But since Russian scientists succeeded in landing a *Venera 7* spacecraft on Venus in 1970, we have known that her crushing atmosphere is dense with poisonous gas and corrosive acid rain.

The planet **Earth** is the third closest planet to the sun, named after the Greek goddess Gaea or Terra Mater, meaning "earth mother." Our planet is the only one in our solar system capable of supporting life as we know it. However other planets—like Mars—may have evolved simple forms of life (e.g., such as bacteria) that became extinct early on in their history. Scientists are anxious to send space probes to the larger moons of Jupiter and Saturn which some scientists suspect may support life of one kind or another. Giant moons like Io, Europa, and Titan may have evolved simple forms of life that are still present in watery oceans beneath their icy surfaces.

The planet **Mars** is the fourth closest planet to the sun, named after the god of "bloody war." Mars's rich red surface is made of iron oxide: simple rust. Mars has the largest volcano in the solar system, Olympus Mons, and also the largest canyon, Valles Marinares. In 1976, American scientists landed *Viking 1* and *Viking 2* on Mars. Both spacecraft sent back millions of bits of information about the planet's surface. Seasonal dust storms on Mars make the hurricane winds of Earth seem like breezes by comparison. The atmosphere on Mars has hardly 1% the atmospheric pressure of Earth, so meteorites can easily make it through to pockmark Mars's surface without being incinerated by friction. One of the most interesting features of the Red Planet is its polar ice caps. During the winter, the martian ice caps—made of water ice and frozen carbon dioxide—can reach latitudes of 40° to 50° from the equator. In summer, the ice caps evaporate and recede to within 10° of the poles.

Beyond the orbit of Mars is the **asteroid belt**. There are more than 5,000 known asteroids swarming along the orbital plane of the planets at average distances of 2.1 to 3.3 **astronomical units** (**AU**) from the Sun. The three largest asteroids are Ceres, Pallas, and Vesta: about 785, 610, and 540 kilometers in diameter, respectively. Asteroids can bump into one another and alter the angle of their orbit. From time to time an errant asteroid collides with one of the planets. Chunks of rock that enter the atmosphere of a planet are called **meteoroids** (e.g., large) or **meteorites** (e.g., small). The planet Jupiter was recently bombarded with the fragments of a giant asteroid that left dark blemishes the size of Earth along the southern latitudes of its clouded atmosphere. The extinction of the dinosaurs on Earth is believed to have resulted from the impact of one such giant rock. The sun is also surrounded by hundreds of frozen chunks of ice called **comets**. Comets revolve around the sun in long elliptical orbits. When a comet nears the sun it starts to vaporize, its glowing tail usually visible from earth.

Homework Directions

You will need index cards, tape, and a spool of string about 60 meters (e.g., 65 yards) long. Start by drawing on an index card a circle 1 centimeter in diameter to represent the sun (e.g., 1 mm = 100,000 kilometers). Tape "sun card" to the end of the string and include a list of basic facts about the sun. As you study this and the following unit on the planets of the solar system, tape other index cards to the string at appropriate distances from the sun using the information you gathered in class. Include drawings of each planet showing the relative size of that planet compared to the sun (e.g., the Earth would be smaller than the period at the end of this sentence). Include facts about each planet as well. At the end of this unit, you should be prepared to present your information about the inner planets and asteroid belt.

Assignment due: _____

_____   _____   ___/___/___
Student's Signature   Parent's Signature   Date

THE INNER PLANETS

Work Date: ____/____/____

LESSON OBJECTIVE

Students will compare and contrast Earth with the planet Mercury.

Classroom Activities

On Your Mark!

Prepare for class by checking out 10–20 reference books containing information about each of the planets of the solar system and the asteroids, comets, and meteors that wander around the sun. Use information in the Teacher's Classwork Agenda and Content Notes to give a brief history of our understanding of the planets and their movements.

Get Set!

Assign groups of cooperative students the task of compiling as much information as they can about a particular planet. Instruct them to use the correct Journal Sheet for the planet they are researching (e.g., Journal Sheet #1 is for information about <u>Mercury</u>, Journal Sheet #2 is for information about <u>Venus</u>, Journal Sheet #3 is for information about <u>Mars</u>, Journal Sheet #4 is for information about the <u>Asteroid Belt</u>).

Go!

Give students the remainder of the period to accomplish the tasks described on the Journal Sheet. Circulate the room adding additional facts that may be of interest to students, asking questions that will help them to visualize what it might be like to make a trip to the planet they are studying

Materials

reference materials, crayons or colored pencils

Name: _____   **Period:** _____   **Date:** ____/____/____

THE INNER PLANETS

MERCURY

Directions: Use the resources provided by your instructor to find and record as much information as you can about the planet Mercury. Draw a picture of this planet to scale with that of Earth making 1 cm = 5,000 km.

approximate diameter
in kilometers:_____

average distance from
the sun in kilometers: _____

average distance from
the sun in astronomical units: _____

period of revolution: _____

period of rotation: _____

number of known satellites: _____

Record additional information on this JOURNAL SHEET about the planet's atmospheric conditions, surface conditions, etc.

THE INNER PLANETS

Work Date: ____/____/____

LESSON OBJECTIVE

Students will compare and contrast Earth with the planet Venus.

Classroom Activities

On Your Mark!

Prepare for class by checking out 10–20 reference books containing information about each of the planets of the solar system and the asteroids, comets, and meteors that wander around the sun. Use information in the Teacher's Classwork Agenda and Content Notes to give a brief history of our understanding of the planets and their movements.

Get Set!

Assign groups of cooperative students the task of compiling as much information as they can about a particular planet. Instruct them to use the correct Journal Sheet for the planet they are researching (e.g., Journal Sheet #1 is for information about <u>Mercury</u>, Journal Sheet #2 is for information about <u>Venus</u>, Journal Sheet #3 is for information about <u>Mars</u>, Journal Sheet #4 is for information about the <u>Asteroid Belt</u>).

Go!

Give students the remainder of the period to accomplish the tasks described on the Journal Sheet. Circulate the room adding additional facts that may be of interest to students, asking questions that will help them to visualize what it might be like to make a trip to the planet they are studying

Materials

reference materials, crayons or colored pencils

EA13 JOURNAL SHEET #2

THE INNER PLANETS

VENUS

Directions: Use the resources provided by your instructor to find and record as much information as you can about the planet Venus. Draw a picture of this planet to scale with that of Earth making 1 cm = 5,000 km.

approximate diameter
in kilometers:_____

average distance from
the sun in kilometers: _____

average distance from
the sun in astronomical units: _____

period of revolution: _____

period of rotation: _____

number of known satellites: _____

Record additional information on this JOURNAL SHEET about the planet's atmospheric conditions, surface conditions, etc.

THE INNER PLANETS

Work Date: ____/____/____

LESSON OBJECTIVE

Students will compare and contrast Earth with the planet Mars.

Classroom Activities

On Your Mark!

Prepare for class by checking out 10–20 reference books containing information about each of the planets of the solar system and the asteroids, comets, and meteors that wander around the sun. Use information in the Teacher's Classwork Agenda and Content Notes to give a brief history of our understanding of the planets and their movements.

Get Set!

Assign groups of cooperative students the task of compiling as much information as they can about a particular planet. Instruct them to use the correct Journal Sheet for the planet they are researching (e.g., Journal Sheet #1 is for information about <u>Mercury</u>, Journal Sheet #2 is for information about <u>Venus</u>, Journal Sheet #3 is for information about <u>Mars</u>, Journal Sheet #4 is for information about the <u>Asteroid Belt</u>).

Go!

Give students the remainder of the period to accomplish the tasks described on the Journal Sheet. Circulate the room adding additional facts that may be of interest to students, asking questions that will help them to visualize what it might be like to make a trip to the planet they are studying

Materials

reference materials, crayons or colored pencils

EA13 Journal Sheet #3

THE INNER PLANETS

MARS

<u>Directions</u>: Use the resources provided by your instructor to find and record as much information as you can about the planet Mars. Draw a picture of this planet to scale with that of Earth making 1 cm = 5,000 km.

approximate diameter
in kilometers:_____

average distance from
the sun in kilometers: _____

average distance from
the sun in astronomical units: _____

period of revolution: _____

period of rotation: _____

number of known satellites: _____

Record additional information on this JOURNAL SHEET about the planet's atmospheric conditions, surface conditions, etc.

THE INNER PLANETS

Work Date: _____/_____/_____

LESSON OBJECTIVE

Students will compare and contrast the characteristics of comets, asteroids, and meteors.

Classroom Activities

On Your Mark!

Prepare for class by checking out 10–20 reference books containing information about each of the planets of the solar system and the asteroids, comets, and meteors that wander around the sun. Use information in the Teacher's Classwork Agenda and Content Notes to give a brief history of our understanding of the planets and their movements.

Get Set!

Assign groups of cooperative students the task of compiling as much information as they can about a particular planet. Instruct them to use the correct Journal Sheet for the planet they are researching (e.g., Journal Sheet #1 is for information about <u>Mercury</u>, Journal Sheet #2 is for information about <u>Venus</u>, Journal Sheet #3 is for information about <u>Mars</u>, Journal Sheet #4 is for information about the <u>Asteroid Belt</u>).

Go!

Give students the remainder of the period to accomplish the tasks described on the Journal Sheet. Circulate the room adding additional facts that may be of interest to students, asking questions that will help them to visualize what it might be like to make a trip to the planet they are studying

Materials

reference materials, crayons or colored pencils

EA13 JOURNAL SHEET #4

THE INNER PLANETS

ASTEROID BELT

<u>Directions</u>: Use the resources provided by your instructor to find and record as much information as you can about the asteroids, comets, and meteors wandering our solar system. Draw pictures comparing and contrasting the nature of asteroids, comets, and meteors. List the names of the discoverers of some of the more well known of these celestial objects.

EA13 REVIEW QUIZ

Directions: Keep your eyes on your own work.
Read all directions and questions carefully.
THINK BEFORE YOU ANSWER!
Watch your spelling, be neat, and do the best you can.

TEACHER'S COMMENTS: _____
_____

THE INNER PLANETS

MULTIPLE CHOICE: Choose the letter of the word or phrase that best completes the sentence or answers the question. *20 points*

_____ 1. Which planet has a nitrogen-oxygen rich atmosphere capable of supporting life?
  (A) Mercury   (D) Mars
  (B) Venus   (E) Ceres
  (C) Earth

_____ 2. Which planet has an atmosphere thick with poisonous gas and corrosive acid rain?
  (A) Mercury   (D) Mars
  (B) Venus   (E) Ceres
  (C) Earth

_____ 3. Which planet has a surface covered with rust?
  (A) Mercury   (D) Mars
  (B) Venus   (E) Ceres
  (C) Earth

_____ 4. Which planet has the shortest year?
  (A) Mercury   (D) Mars
  (B) Venus   (E) Ceres
  (C) Earth

_____ 5. Which planet has the hottest surface temperature?
  (A) Mercury   (D) Mars
  (B) Venus   (E) Ceres
  (C) Earth

_____ 6. Which planet has ice caps made of water ice and frozen carbon dioxide?
   (A) Mercury   (D) Mars
   (B) Venus   (E) Ceres
   (C) Earth

_____ 7. Which is not a planet?
   (A) Mercury   (D) Mars
   (B) Venus   (E) Ceres
   (C) Earth

_____ 8. Which planet has a day that is longer than its year?
   (A) Mercury   (D) Mars
   (B) Venus   (E) Ceres
   (C) Earth

_____ 9. Which planet has not been the object of a landing by a space probe?
   (A) Mercury   (D) Mars
   (B) Venus   (E) Ceres
   (C) Earth

_____ 10. Which planet has two moons?
   (A) Mercury   (D) Mars
   (B) Venus   (E) Ceres
   (C) Earth

DIAGRAM: Use arrows and labels to identify the shaded area and the four circles by their name. *10 points*

sun

THE OUTER PLANETS

TEACHER'S CLASSWORK AGENDA AND CONTENT NOTES

Classwork Agenda for the Week

1. Students will compare and contrast Earth with the planet Jupiter.
2. Students will compare and contrast Earth with the planet Saturn.
3. Students will compare and contrast Earth with the planets Uranus and Neptune.
4. Students will compare and contrast Earth with the planet Pluto.

Content Notes for Lecture and Discussion

The purpose of any good theory is to explain a number of phenomena in a simple and concise fashion and make predictions about future discoveries. It is the goal of science to find exceptions to contemporary theories in order to refine our understanding of the universe by making a theory as complete as it can be. Any theory whose predictions fail to be sufficiently accurate must be painstakingly refined or summarily discarded. One of the most famous theories to drive astronomical research for more than a century was the **Titius-Bode law**. According to the Titius-Bode law, derived by German astronomers **Johann Daniel Titius** (b. 1747; d. 1796) and **Johann Elert Bode** (b. 1747; d. 1826), there is a definite numerical sequence giving the distances of the planets from the sun. Putting the Earth at the arbitrary distance of 10 units from the sun, the distances of Mercury, Venus, Earth, Jupiter, and Saturn become 4, 4+3, 4+6, 4+12, 4+48, and 4+96, respectively. The missing number in the sequence, 4+24, represented by a missing planet between Mars and Jupiter which was later found to be the asteroid belt. The discovery of Uranus at 19.2 A.U. from the sun seemed in good agreement with the law, which predicted a planet to be 4+192 in the sequence. Neptune and Pluto, however, were exceptions to the law which forced the abandonment of the theory.

The planets Jupiter and Saturn are the only two outer planets visible to the naked eye. The planet Uranus was discovered by English astronomer **Sir William Herschel** (b. 1738; d. 1822) in

planet	Jupiter	Saturn	Uranus	Neptune	Pluto
avg. distance to sun (10^6 km)	778.3	1,429	2,875	4,504	5.900
orbital eccentricity	0.05	0.06	0.05	0.01	0.25
orbital inclination	1.3°	2.5°	0.8°	1.8°	17.2°
period of revolution	11.86 years	29.46 years	84 years	165 years	248 years
period of rotation	9.92 hours	10.66 hours	17.3 hours	17.83 hours	6.39 days
orbital velocity (km/sec)	13	9.6	6.8	5.4	4.7
diameter at equator (km)	142,948	120,536	51,100	49,200	3,200
volume compared to earth	1,319	735	67	57	0.1
mass compared to earth	317.9	95.2	14.54	17.2	0.002
density compared to water	1.3	0.7	1.2	1.56	0.8
gravity compared to earth	2.53	1.07	0.91	1.16	0.05
escape velocity	60.3 km/s	36.3 km/s	22.6 km/s	25 km/s	5.2 km/s
albedo	0.73	0.76	0.93	0.84	unknown
known satellites	16	17	15	2	1

OUTER PLANET BASIC FACTS

1781. Herschel used a simple 7-inch reflecting telescope that he had built himself. The existence of the planet Neptune was predicted by the English astronomer **John Couch Adams** (b. 1819; d. 1892) in 1845 based on his observations of the variation in the orbit of Uranus. He deduced that another planet—later found to be Neptune—was exerting a strong gravitational attraction that altered Uranus's path. Neptune was discovered in the following year by the German astronomer **Johann Gottfried Galle** (b. 1812; d. 1910). The existence of Pluto was predicted using a similar argument. American astronomer **Percival Lowell** (b. 1855; d. 1916) deduced the existence of Pluto by comparing variations in the orbits of both Uranus and Neptune. American astronomers **William Henry Pickering** (b. 1858; d. 1938) and **Clyde William Tombaugh** (b. 1906; d. 1997) finally sighted Pluto in 1930. Charon, Pluto's only moon, was discovered in 1978. Charon completes one revolution of Pluto every 6.39 days, the same time it takes Pluto to rotate once on its axis.

In Lesson #1, students will compare and contrast Earth with the planet Jupiter.

In Lesson #2, students will compare and contrast Earth with the planet Saturn.

In Lesson #3, students will compare and contrast Earth with the planets Uranus and Neptune.

In Lesson #4, students will compare and contrast Earth with the planet Pluto.

ANSWERS TO THE HOMEWORK PROBLEMS

Students complete their string of the inner and outer planets by the end of this unit. As 1 millimeter is equal to 100,000 kilometers, on the homework scale 1 centimeter equals 1,000,000 kilometers. The index card for the planet Jupiter should be placed 7.78 meters from the sun; Saturn should be placed 14.29 meters from the sun; Uranus, 28.75 m; Neptune 45.04 m; and, Pluto 59.0 m from the sun. Point out that the orbital path of Pluto is highly elliptical which has placed it closer to the sun than Neptune. It will move farther from the sun than Neptune within the next decade.

ANSWERS TO THE END-OF-THE-WEEK REVIEW QUIZ

1. E
2. A
3. B
4. C
5. D

6. A
7. E
8. A
9. E
10. B

EA14 Fact Sheet

THE OUTER PLANETS

CLASSWORK AGENDA FOR THE WEEK

(1) Compare and contrast the Earth with the planet Jupiter.
(2) Compare and contrast the Earth with the planet Saturn.
(3) Compare and contrast the Earth with the planets Uranus and Neptune.
(4) Compare and contrast the Earth with the planet Pluto.

The planets Jupiter, Saturn, Uranus, Neptune and Pluto comprise **the outer planets** of the solar system. The first four planets listed are enormous gaseous worlds that could swallow over 1,000 earths. The last, the planet Pluto, is a cold, icy rock barely one-third the diameter of our planet.

The planet **Jupiter** is the fifth planet from the sun, named for the "king of the gods." It is the largest planet, colorfully banded with bright streaks of red and orange clouds that move across its stormy latitudes. A Great Red Spot adorns its southern hemisphere: a swirling hurricane three times the diameter of Earth. In 1979, the space probe *Voyager 1* flew by Jupiter taking pictures of its many moons and atmosphere. It is believed that Jupiter may have a small, dense molten core surrounded by frozen water and ammonia or even metallic hydrogen. In 1995, the *Galileo* space probe passed close enough to Jupiter to send a smaller probe down to the planet. *Galileo's* tiny probe survived for nearly an hour, sending back important information about the planet's upper atmosphere.

The planet **Saturn** is the second largest planet, located sixth from the sun. Saturn is named for the god of "the harvest." It is the only planet with an extensive ring system visible from Earth with a simple telescope. **Galileo Galilei** (b. 1564; d. 1642) was the first to see Saturn's rings in 1610. The other gaseous planets have smaller ring systems much more difficult to observe from a great distance. *Voyager 1* passed by Saturn in 1980 before continuing its journey out of our solar system into **interstellar space**. *Voyager 1* took pictures of the thousands of individual "ringlets" that make up the rings encircling the planet. Like Jupiter, Saturn has a number of giant moons: Titan—the largest—is larger than the planets Mercury or Pluto. Despite its size Saturn is less dense than liquid water. If the solar system were a giant ocean made of water, the planet Saturn would float!

The planet **Uranus** is the seventh planet from the sun, named for the father of Saturn. It is a giant blue-green marble less than half the diameter of Saturn, discovered by the English astronomer **William Frederick Herschel** (b. 1738; d. 1822) in 1781. Its density is slightly more than that of water, but the gravitational attraction at its surface is less than that of earth. Perhaps the most striking feature of Uranus is its tilt. While the other planets are tilted only slightly off the perpendicular to the ecliptic—like Earth at about 23.5°—Uranus is tipped completely on its side! The planet's South Pole—not its equator—faces toward the sun. Uranus is tilted more than 82° off the perpendicular to the ecliptic. *Voyager 2* made a trip past Uranus in 1986, discovering its ringlets and taking samples of its azure methane atmosphere.

The planet **Neptune** is normally the eighth planet from the sun but at present it is farther from the sun than Pluto. The reason for this is the "eccentricity" of Pluto's orbit discussed in the next paragraph. Neptune is named for the god of the "sea." *Voyager 2* flew by Neptune on its way out of the solar system in 1989. Neptune resembles Uranus in size and atmospheric composition but is not tilted on its side like its strange neighbor. While Neptune was discovered by the German astronomer **Johann Gottfried Galle** (b. 1812; d. 1910) in 1846 its discovery was predicted by the English astronomer **John Couch Adams** (b. 1819; d. 1892) in 1845. Adams explained the variation in the orbit of Uranus by suggesting that another planet's gravity—later found to be Neptune—was tugging Uranus off its expected orbital path.

The planet **Pluto** is the ninth planet of the solar system, named for the god of the "underworld." It was discovered in 1930 by the American astronomers **William Henry Pickering** (b. 1858; d. 1938) and **Clyde William Tombaugh** (b. 1906; d. 1997). Like Neptune, the existence of Pluto was deduced by comparing variations in the orbits of other planets (e.g., Uranus and Neptune). The prediction was made by the American astronomer **Percival Lowell** (b. 1855; d. 1916) decades before Pluto was actually sighted. Although little is known about the surface and atmosphere of Pluto—as it has yet to be visited by a space probe—it is known to have the most elliptical orbit of all the planets. The eccentricity of Pluto's orbit, and the fact that it orbits the sun at an angle of 17° off the ecliptic, account for Pluto's being closer to the sun at present than the planet Neptune. Even more remarkable is the fact that Pluto's moon **Charon** orbits at a distance of only 19,000 kilometers. Being more than one-third the diameter of Pluto, Charon would cover nearly one-quarter of Pluto's sky when viewed from the planet's surface. Named for the "ferryman" who takes the dead across the river Styx to the underworld, Charon is more than a moon to Pluto. Together they actually comprise a "double planet."

Homework Directions

Continue the homework assignment you began in the previous unit. As you study the outer planets of the solar system, tape index cards to the string at appropriate distances from the sun. Include drawings of each planet showing the relative size and distance of that planet compared to the sun (e.g., 1 mm = 100,000 kilometers). Include facts about each planet as well. At the end of this unit, you should be prepared to present your information about all of the planets of the solar system.

Assignment due: _____

_____   _____   ____/____/____
Student's Signature   Parent's Signature   Date

THE OUTER PLANETS

Work Date: ____/____/____

LESSON OBJECTIVE

Students will compare and contrast Earth with the planet Jupiter.

Classroom Activities

On Your Mark!

Prepare for class by checking out 10–20 reference books containing information about each of the planets of the solar system and the asteroids, comets, and meteors that wander around the sun. Use information in the Teacher's Classwork Agenda and Content Notes to give a brief history of our understanding of the planets and their movements.

Get Set!

Assign groups of cooperative students the task of compiling as much information as they can about a particular planet. Instruct them to use the correct Journal Sheet for the planet they are researching (e.g., Journal Sheet #1 is for information about <u>Jupiter</u>, Journal Sheet #2 is for information about <u>Saturn</u>, Journal Sheet #3 is for information about <u>Uranus</u> and <u>Neptune</u>, Journal Sheet #4 is for information about <u>Pluto</u>).

Go!

Give students the remainder of the period to accomplish the tasks described on the Journal Sheet. Circulate the room adding additional facts that may be of interest to students, asking questions that will help them to visualize what it might be like to make a trip to the planet they are studying

Materials

reference materials, crayons or colored pencils

EA14 JOURNAL SHEET #1

THE OUTER PLANETS

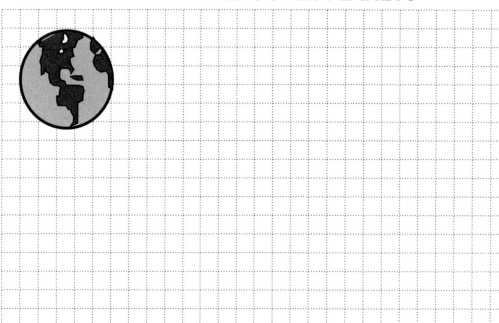

JUPITER

<u>Directions</u>: (1) Use the resources provided by your instructor to find and record as much information as you can about the planet Jupiter. (2) Use a piece of string to show the radius of this planet so that 1 cm = 5,000 km. (3) Use the string like a compass to draw a curved section of the planet's surface on this JOURNAL SHEET. (4) Compare the curved section to the size of Earth shown above which is pictured using the same scale.

approximate diameter
in kilometers:_____

average distance from
the sun in kilometers: _____

average distance from
the sun in astronomical units: _____

period of revolution: _____

period of rotation: _____

number of known satellites: _____

Record additional information on this JOURNAL SHEET about the planet's atmospheric conditions, surface conditions, etc.

THE OUTER PLANETS

Work Date: ____/____/____

LESSON OBJECTIVE

Students will compare and contrast Earth with the planet Saturn.

Classroom Activities

On Your Mark!

Prepare for class by checking out 10–20 reference books containing information about each of the planets of the solar system and the asteroids, comets, and meteors that wander around the sun. Use information in the Teacher's Classwork Agenda and Content Notes to give a brief history of our understanding of the planets and their movements.

Get Set!

Assign groups of cooperative students the task of compiling as much information as they can about a particular planet. Instruct them to use the correct Journal Sheet for the planet they are researching (e.g., Journal Sheet #1 is for information about <u>Jupiter</u>, Journal Sheet #2 is for information about <u>Saturn</u>, Journal Sheet #3 is for information about <u>Uranus</u> and <u>Neptune</u>, Journal Sheet #4 is for information about <u>Pluto</u>).

Go!

Give students the remainder of the period to accomplish the tasks described on the Journal Sheet. Circulate the room adding additional facts that may be of interest to students, asking questions that will help them to visualize what it might be like to make a trip to the planet they are studying

Materials

reference materials, crayons or colored pencils

Name: _____ Period: _____ Date: ____/____/____

EA14 Journal Sheet #2

THE OUTER PLANETS

SATURN

Directions: (1) Use the resources provided by your instructor to find and
record as much information as you
can about the planet Saturn. (2) Use a
piece of string to show the radius of
this planet so that 1 cm = 5,000 km. (3)
Use the string like a compass to draw
a curved section of the planet's surface
on this JOURNAL SHEET. (4)
Compare the curved section to the
size of Earth shown above which is
pictured using the same scale.

approximate diameter
in kilometers:_____

average distance from
the sun in kilometers: _____

average distance from
the sun in astronomical units: _____

period of revolution: _____

period of rotation: _____

number of known satellites: _____

Record additional information on this
JOURNAL SHEET about the planet's
atmospheric conditions, surface
conditions, etc.

190

THE OUTER PLANETS

Work Date: ____/____/____

LESSON OBJECTIVE

Students will compare and contrast Earth with the planets Uranus and Neptune.

Classroom Activities

On Your Mark!

Prepare for class by checking out 10–20 reference books containing information about each of the planets of the solar system and the asteroids, comets, and meteors that wander around the sun. Use information in the Teacher's Classwork Agenda and Content Notes to give a brief history of our understanding of the planets and their movements.

Get Set!

Assign groups of cooperative students the task of compiling as much information as they can about a particular planet. Instruct them to use the correct Journal Sheet for the planet they are researching (e.g., Journal Sheet #1 is for information about Jupiter, Journal Sheet #2 is for information about Saturn, Journal Sheet #3 is for information about Uranus and Neptune, Journal Sheet #4 is for information about Pluto).

Go!

Give students the remainder of the period to accomplish the tasks described on the Journal Sheet. Circulate the room adding additional facts that may be of interest to students, asking questions that will help them to visualize what it might be like to make a trip to the planet they are studying

Materials

reference materials, crayons or colored pencils

Name: _____   Period:_____   Date: ____/____/____

EA14 JOURNAL SHEET #3

THE OUTER PLANETS

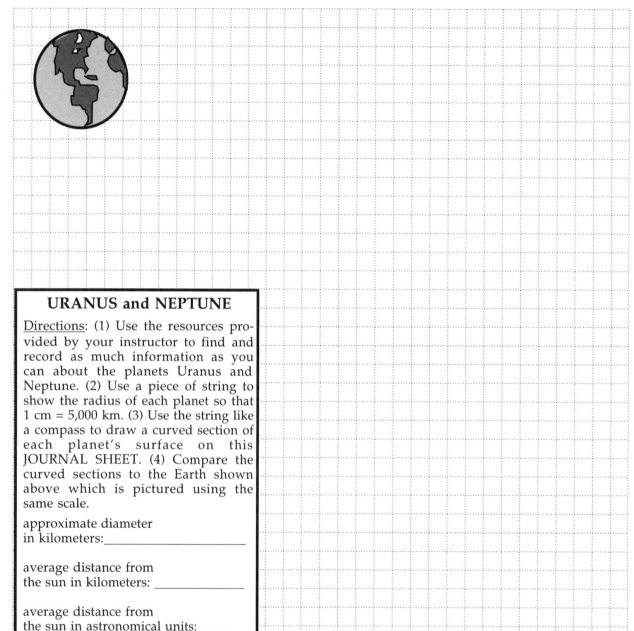

URANUS and NEPTUNE

Directions: (1) Use the resources provided by your instructor to find and record as much information as you can about the planets Uranus and Neptune. (2) Use a piece of string to show the radius of each planet so that 1 cm = 5,000 km. (3) Use the string like a compass to draw a curved section of each planet's surface on this JOURNAL SHEET. (4) Compare the curved sections to the Earth shown above which is pictured using the same scale.

approximate diameter
in kilometers:_____

average distance from
the sun in kilometers: _____

average distance from
the sun in astronomical units: _____

period of revolution: _____

period of rotation: _____

number of known satellites: _____

Record additional information on this JOURNAL SHEET about the planet's atmospheric conditions, surface conditions, etc.

EA14 Lesson #4

THE OUTER PLANETS

Work Date: ____/____/____

LESSON OBJECTIVE

Students will compare and contrast Earth with the planet Pluto.

Classroom Activities

On Your Mark!

Prepare for class by checking out 10–20 reference books containing information about each of the planets of the solar system and the asteroids, comets, and meteors that wander around the sun. Use information in the Teacher's Classwork Agenda and Content Notes to give a brief history of our understanding of the planets and their movements.

Get Set!

Assign groups of cooperative students the task of compiling as much information as they can about a particular planet. Instruct them to use the correct Journal Sheet for the planet they are researching (e.g., Journal Sheet #1 is for information about <u>Jupiter</u>, Journal Sheet #2 is for information about <u>Saturn</u>, Journal Sheet #3 is for information about <u>Uranus</u> and <u>Neptune</u>, Journal Sheet #4 is for information about <u>Pluto</u>).

Go!

Give students the remainder of the period to accomplish the tasks described on the Journal Sheet. Circulate the room adding additional facts that may be of interest to students, asking questions that will help them to visualize what it might be like to make a trip to the planet they are studying

Materials

reference materials, crayons or colored pencils

EA14 JOURNAL SHEET #4

THE OUTER PLANETS

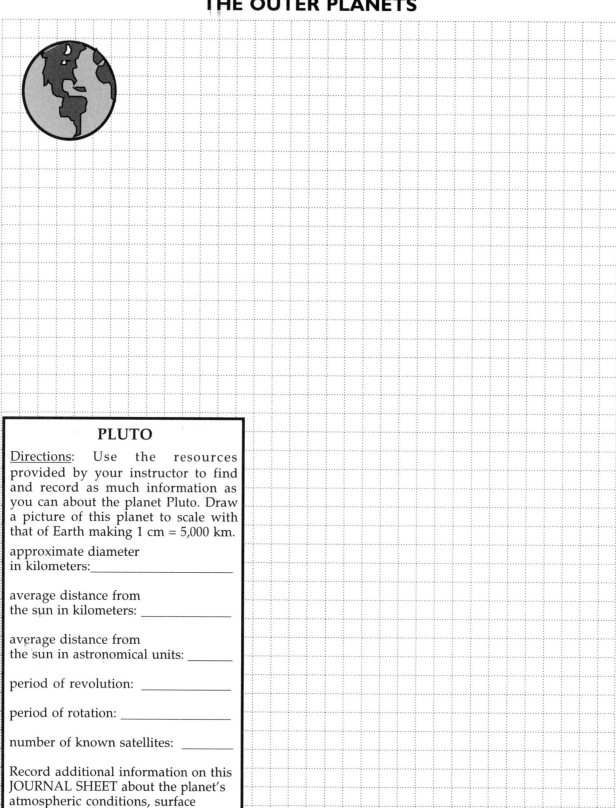

PLUTO

<u>Directions</u>: Use the resources provided by your instructor to find and record as much information as you can about the planet Pluto. Draw a picture of this planet to scale with that of Earth making 1 cm = 5,000 km.

approximate diameter
in kilometers:_____

average distance from
the sun in kilometers: _____

average distance from
the sun in astronomical units: _____

period of revolution: _____

period of rotation: _____

number of known satellites: _____

Record additional information on this JOURNAL SHEET about the planet's atmospheric conditions, surface conditions, etc.

EA14 REVIEW QUIZ

Directions: Keep your eyes on your own work.
Read all directions and questions carefully.
THINK BEFORE YOU ANSWER!
Watch your spelling, be neat, and do the best you can.

CLASSWORK   (~40): _____
HOMEWORK   (~20): _____
CURRENT EVENT (~10): _____
TEST   (~30): _____

TOTAL   (~100): _____
(A ≥ 90, B ≥ 80, C ≥ 70, D ≥ 60, F < 60)

LETTER GRADE: _____

TEACHER'S COMMENTS: _____
_____

THE OUTER PLANETS

MULTIPLE CHOICE: Choose the letter of the word or phrase that best completes the sentence or answers the question. *20 points*

_____ 1. Which planet is actually a "double planet"?
      (A) Jupiter   (D) Neptune
      (B) Saturn   (E) Pluto
      (C) Uranus

_____ 2. Which planet was recently bombarded by fragments from a giant asteroid?
      (A) Jupiter   (D) Neptune
      (B) Saturn   (E) Pluto
      (C) Uranus

_____ 3. Which planet has the most visible rings from earth?
      (A) Jupiter   (D) Neptune
      (B) Saturn   (E) Pluto
      (C) Uranus

_____ 4. Which planet is tipped almost completely on its side?
      (A) Jupiter   (D) Neptune
      (B) Saturn   (E) Pluto
      (C) Uranus

_____ 5. Which planet was the first to be discovered after its existence was predicted?
      (A) Jupiter   (D) Neptune
      (B) Saturn   (E) Pluto
      (C) Uranus

_____ 6. Which planet has a hurricane large enough to swallow three earths?
    (A) Jupiter   (D) Neptune
    (B) Saturn   (E) Pluto
    (C) Uranus

_____ 7. Which planet has the closest moon?
    (A) Jupiter   (D) Neptune
    (B) Saturn   (E) Pluto
    (C) Uranus

_____ 8. Which planet is the largest?
    (A) Jupiter   (D) Neptune
    (B) Saturn   (E) Pluto
    (C) Uranus

_____ 9. Which planet is the only planet that definitely has a solid surface?
    (A) Jupiter   (D) Neptune
    (B) Saturn   (E) Pluto
    (C) Uranus

_____ 10. Which planet is the least dense?
    (A) Jupiter   (D) Neptune
    (B) Saturn   (E) Pluto
    (C) Uranus

DIAGRAM: Use arrows and labels to identify the shaded area and the nine circles by their name. *10 points*

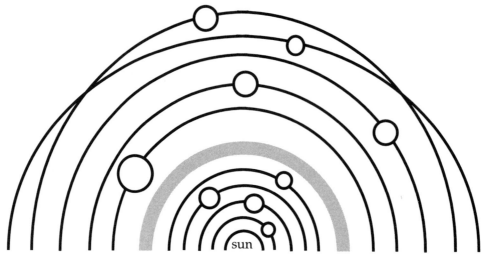

_____   _____   ___/___/___
Student's Signature   Parent's Signature   Date

EA15

STARS AND GALAXIES
OF THE COSMOS

TEACHER'S CLASSWORK AGENDA AND CONTENT NOTES

Classwork Agenda for the Week

1. Students will describe the methods scientists use to measure the distance to faraway stars.

2. Students will describe the forces that give birth to stars and galaxies.

3. Students will describe the forces that cause stars to age and die.

4. Students will examine evidence that the universe is expanding.

Content Notes for Lecture and Discussion

By the end of the 19th century the arts of telescopy, spectroscopy, and photography had expanded the horizons of the universe. In 1912, American astronomer **Henrietta Swan Leavitt** (b. 1868; d. 1921) discovered a relationship between the brightness of a class of stars called **Cepheid variables** and their rates of pulsation (e.g., periodically changing luminosity). Leavitt's **period-luminosity law** allowed astronomers to use the standard **law of luminosity** to measure the distance to these stars. The law of luminosity states that the brightness of a star diminishes with the square of its distance from an observer. What was more exciting was the fact that Leavitt's rule could be used to calculate the distance to faraway galaxies that contain Cepheid variables like our own Milky Way. Danish astronomer **Ejnar Hertzsprung** (b. 1873; d. 1957) and American astronomer **Henry Norris Russell** (b. 1877; d. 1957) used Leavitt's and other spectroscopic data to make further determinations of stellar composition and stellar luminosity. This allowed astronomers to achieve more accurate estimations of stellar distances and to suggest a possible mechanism for the evolution of the stars. **Albert Einstein**'s (b. 1879; d. 1955) formulation, $E = mc^2$, allowed physicists to comprehend how stars evolved in a series of stages that pitted the forces of nuclear fusion and gravity against one another. Using the **Hertzsprung-Russell diagram** American physicist **Edwin Powell Hubble** (b. 1889; d. 1953) was able to prove in 1924 that galaxies were groups of stars outside our own Milky Way and to measure their distances and rates of recession. This information led to his conclusion that the universe is expanding as Einstein had predicted in his publication of his **general theory of relativity** in 1916.

In 1927, the Belgian astronomer **Georges Edouard Lemaître** (b. 1894; d. 1966) proposed the **Big Bang Theory** later popularized by American physicist **George Gamow** (b. 1904; d. 1968). Since then, cosmologists have contemplated the fate of the universe: Will it expand forever? The rate of expansion can be measured using the data gathered by Hubble and others and the degree of gravitational attraction needed to halt the expansion deduced from this information. If the universe has sufficient density, then its expansion will cease and the "big crunch" will begin. The density of matter needed to halt the expansion is on the order of 3 protons per cubic meter of space. Actual observations show the universe to be 90% short of the required mass. As a result of these findings, the search for elusive **dark matter** goes on.

In Lesson #1, students will describe the methods scientists use to measure the distance to faraway stars.

In Lesson #2, students will describe the forces that give birth to stars and galaxies.

In Lesson #3, students will describe the forces that cause stars to age and die.

In Lesson #4, students will examine evidence that the universe is expanding.

EA15 Content Notes (cont'd)

ANSWERS TO THE HOMEWORK PROBLEMS

Students will discover that their hand "wobbles" as they try to keep the sock revolving. Scientists see stars wobbling in this manner all the time. The wobble suggests that the star is being pulled this way and that by another massive object. Since the other object is difficult to see because it is not emitting light, scientists conclude that it might be a planet in orbit around the wobbling star.

ANSWERS TO THE END-OF-THE-WEEK REVIEW QUIZ

1. cosmology	6. fusion	11. G	16. F
2. expanding	7. matter or gravity	12. C	17. D
3. Big Bang	8. true	13. I	18. E
4. true	9. true	14. H	19. A
5. hydrogen	10. billion	15. J	20. B

EA15 FACT SHEET

STARS AND GALAXIES OF THE COSMOS

CLASSWORK AGENDA FOR THE WEEK

(1)   Describe the methods scientists use to measure the distance to faraway stars.
(2)   Describe the forces that give birth to stars and galaxies.
(3)   Describe the forces that cause stars to age and die.
(4)   Examine evidence that the universe is expanding.

Cosmology is the study of the universe. It includes theories about how the universe began, how it is evolving, and what might happen to it in the future.

In the 1930s the American physicist Edwin Powell Hubble (b. 1889; d. 1953) observed that all the known galaxies are moving away from one another. He suggested that the universe is expanding. In 1948, the American physicist George Gamow (b. 1904; d. 1968) imagined what it would be like if you could watch the galaxies rushing toward one another—the reverse of what is actually happening—as you took a journey back in time. Arriving at the beginning of time, you would see all the matter in the universe squeezed by enormous gravitational forces into a space much smaller than an atom. At that point, the universe would have no volume and infinite density. This theoretical creation is called a space-time singularity. Gamow proposed that the universe began about 15 billion years ago when such a singularity exploded. The explosion created space and all the matter in the universe throwing it outward into the void. Since Gamow's proposal, the idea has become known as the Big Bang Theory.

Shortly after the universe began, the element hydrogen was formed. Hydrogen makes up more than 90% of the known universe and is the primary fuel burned by new young stars like our sun. Gravity pulls hydrogen atoms together until they fuse and release enormous amounts of energy. A star is born when the energy released by this fusion process balances the force of gravity pulling in the hydrogen atoms. When a star's hydrogen fuel is exhausted the star's gravity causes the star to collapse. As it collapses its center becomes hotter and hotter until nuclear fusion begins again. This time the star will burn the heavier elements manufactured in its youth. Pressure inside the star caused by nuclear fusion forces the star to expand into a red giant star. When a red giant star exhausts its nuclear fuel it will also collapse, then explode. An exploding star is called a nova or supernova. The future of the star following this violent collapse and nova depends on the amount of mass it had when it was born. A star like our own sun will eventually form a white dwarf star. A star that is more than 1.5 times the mass of our sun will form a neutron star. A star 2.5 times the mass of our own sun will eventually collapse completely into a mysterious black hole. There are billions of stars in the universe that will do this in the future. Stars live for billions of years. Our own sun is about 5 billion years old and will not become a red giant for another 5 billion years.

Galaxies are composed of billions of stars. Stars form galaxies of many different types: spiral galaxies, bar-shaped galaxies, and globular galaxies. Galaxies form galactic clusters and galactic clusters form super clusters of galaxies. Our own Milky Way Galaxy is a spiral galaxy believed to be about 70,000 light-years in diameter. However, there are many smaller and larger galaxies spanning the cosmos. The best estimates of the size of the universe are between 10 and 20 billion light-years across.

The universe continues to expand today and may someday stop expanding and begin to collapse. Whether or not it does collapse depends on how much matter and gravity there is in the cosmos as a whole. If there is sufficient matter in the universe, then the force of gravity will eventually stop the expansion and pull the whole universe back together again. The universe will contract until another singularity is formed. Scientists wonder what would happen after our universe collapsed into a singularity. Would it explode again with another "big bang" and create a whole new universe?

Homework Directions

Perform the following activity: (1) Stuff an old sock with a ball. (2) Tie a meter-long string to the sock so that the ball cannot spill out. (3) Swing the sock at the end of the string over your head noting the motion of your hand and the sock. (4) Draw the movement made by your hand as the sock revolves around it. (5) Considering your observations of your moving hand and the revolving sock, explain how scientists use the "wobble" of distant stars to try and detect planets that might be revolving around them.

Assignment due: _____

_____   _____   ____/____/____
Student's Signature   Parent's Signature   Date

STARS AND GALAXIES OF THE COSMOS

Work Date: ____/____/____

LESSON OBJECTIVE

Students will describe the methods scientists use to measure the distance to faraway stars.

Classroom Activities

On Your Mark!

Review the concept of **parallax** as introduced in Lesson #4 of Unit #10: *Mapping the Heavens*. Explain that parallax is an effective method for measuring the distance to stars that are close by (e.g., within 300 light years). When dividing the night sky into degrees azimuth and declination a single degree covers a large area. One degree can be further divided into 60 **minutes of arc**. Each minute of arc equals 60 **seconds of arc**. When a star has a parallax of 1 second of arc it is **1 parsec** away from the sun. One parsec is 3.26 light-years. The closest star to our sun—Alpha Centauri—is 4 light-years away (e.g., 9 trillion kilometers)! The distances to stars that are farther away must be estimated using other methods.

Get Set!

Ancient astronomers estimated the distances to the stars by their brightness or **apparent magnitude**, since it was thought that all stars were like our sun and burned with equal brightness. However, the apparent magnitude of a star or distant galaxy depends on the intrinsic brightness of the star (e.g., its energy output) and its distance from the observer. Hold a glass prism up to the light of a white light bulb to show how light refracts or bends into the separate colors of the **spectrum** (e.g., red, orange, yellow, green, blue, indigo, violet). Explain that every chemical element burns at high temperatures giving off a specific color or spectral "fingerprint." Point out that astronomers use the spectral fingerprints of stars to deduce the elements burning at or near the star's surface. The intrinsic brightness of a star can be estimated by photographing and analyzing the star's spectrum. The analysis gives clues to the surface temperature of the star which can be used to estimate the star's total energy output or **luminosity**. The relationship between a star's surface temperature and luminosity can be graphed on the **Hertzsprung-Russell diagram** shown on Figure A on Journal Sheet #1. The diagram was devised by the Danish astronomer **Ejnar Hertzsprung** (b. 1873; d. 1957) and the Amercian astronomer **Henry Norris Russell** (b. 1877; d. 1957). Knowing its energy output allows scientists to estimate its distance according to the **law of illumination**. The law states that the amount of illumination decreases as the square of the distance from the source. Given two stars—"a" and "b"—of equal size and surface composition, star "b" burning with one-quarter the brightness of star "a," one can estimate star "b" to be twice the distance away as star "a." A star twice as far away as a known star of equal luminosity will shine with one-quarter the known star's brightness.

Go!

Give students ample time to perform the activity described in Figure B on Journal Sheet #1. They will find that the number of lit squares increases as the distance to the last card increases. At the same time, the brightness of the light on the card becomes dimmer.

Materials

light bulbs, dark construction paper, scissors, prisms

EA15 JOURNAL SHEET #1

STARS AND GALAXIES OF THE COSMOS

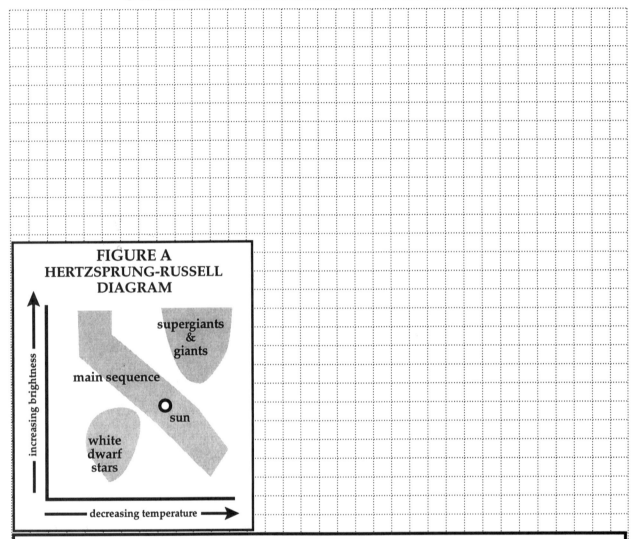

FIGURE A
HERTZSPRUNG-RUSSELL DIAGRAM

increasing brightness

supergiants & giants

main sequence

O sun

white dwarf stars

decreasing temperature

FIGURE B

<u>Directions</u>: (1) Cut out three pieces of dark construction paper 15 cm by 20 centimeters. (2) Mark off the horizontal and vertical axes of the construction paper at 1 cm intervals. (3) Draw a grid of small 1 cm x 1 cm squares on each piece of paper creating a grid that is 15 cm by 15 cm as shown. (4) Cut the bottom of the paper as shown to make "feet" so that each piece can stand on its own. (5) Cut out the center square in the first piece of paper. (6) Cut out the center 25 squares of the second piece. (7) And <u>draw</u> a 9 x 9 square on the third piece. (8) Darken the room and place the first piece close to a lighted 40 watt bulb; the second piece 30 cm away; and, the third 60 cm away. NOTE: The drawing shown here is not to scale! (9) Count the number of squares that are illuminated on the third piece of paper. Move the last piece to 90 cm away and count the number of squares again. Write a conclusion that describes what happens to the intensity of light as you move away from the source.

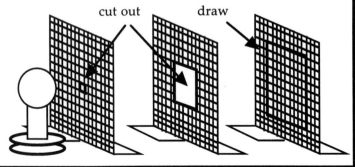

cut out   draw

STARS AND GALAXIES OF THE COSMOS

Work Date: ____/____/____

LESSON OBJECTIVE

Students will describe the forces that give birth to stars and galaxies.

Classroom Activities

On Your Mark!

Prepare for class by checking out 10–20 reference books containing information about the different kinds of stars in the universe (e.g., **yellow stars** like our sun, **red giants** and **supergiants**, **white dwarfs**, etc.).

Review the current theory regarding the formation of the solar system introduced in Lesson #1, Unit 5: *Earth Origins and Geologic Time*. Remind students that the force of attraction between every object in the universe called gravity pervades the universe. Explain that stars are formed out of clouds of dust and gas called **nebulae**. Some nebulae are the remains of old stars that have exploded, throwing out much of their matter into space. Gravity pulls the material of nebulae together into a rotating disk. At the center of the disk, where gravity is the strongest, the temperature rises to millions of degrees. Atoms begin to collide and fuse together. Atoms of hydrogen fuse together to form atoms of helium. Helium atoms can fuse together to form atoms of carbon, nitrogen, and some of the heavier elements.

Get Set!

Nuclear fusion releases an enormous amount of energy, creating the "internal pressure" that prevents the young star from completely collapsing. Introduce **Albert Einstein**'s (b. 1879; d. 1955) famous equation: $E = mc^2$. Explain that "E" stands for "energy"; "m" stands for "mass"; and "c" stands for the "speed of light" (e.g., 300,000,000 meters per second). Ask students to guess how much energy would be released if we could fuse together all of the hydrogen atoms in a glass of water (e.g., about 1 kilogram of hydrogen). Have them copy your calculations on Journal Sheet #2.

E	=	m	×	c^2		
	=	m	×	c	×	c
	=	1 kg	×	300,000,000 m/s	×	300,000,000 m/s
	=	90,000,000,000,000,000 joules of energy				
	=	90 million billion joules of energy				

Point out that a 40-watt light bulb burns 40 joules of energy every second. So, the energy released from the nuclear fusion of 1 kg of hydrogen could burn 2.25 million billion 40-watt light bulbs for 1 second; or, 625,000,000,000 40-watt light bulbs for 1 hour; or, 26,041,000,000 40-watt light bulbs for a day; or, 71,000,000 40-watt light bulbs for a year. Mention that our sun fuses trillions of kilograms of hydrogen every second! This amount of energy "output" works against gravity to keep the star from collapsing. When the star exhausts its fuel, gravity takes over and the star collapses until nuclear fusion ignites to burn some of the heavier elements created in the younger star. This collapse and ignition is associated with a great explosion called a **nova** that throws off dust and debris.

Go!

Give students ample time to gather information about the different types of stars described in the reference materials you provide.

Materials

library references

EA15 JOURNAL SHEET #2

STARS AND GALAXIES OF THE COSMOS

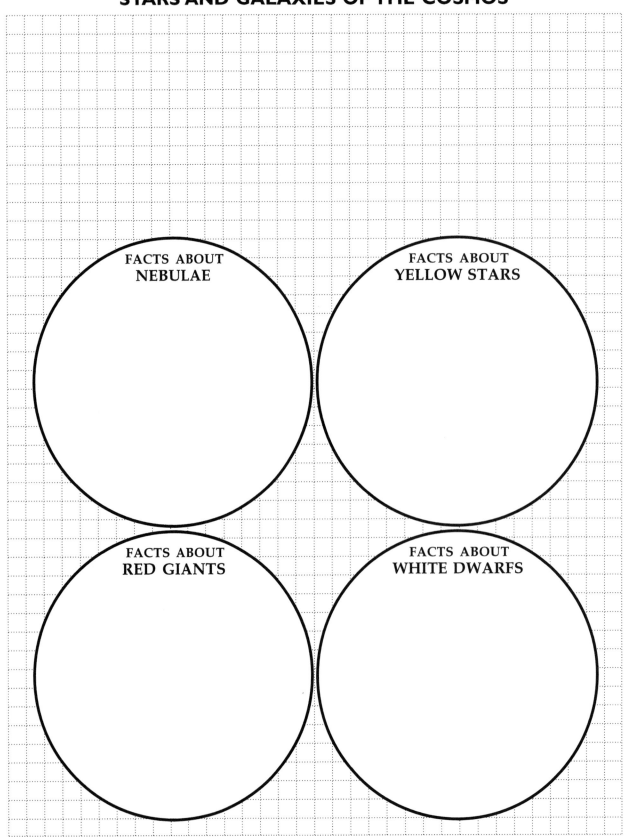

FACTS ABOUT
NEBULAE

FACTS ABOUT
YELLOW STARS

FACTS ABOUT
RED GIANTS

FACTS ABOUT
WHITE DWARFS

STARS AND GALAXIES OF THE COSMOS

Work Date: ____/____/____

LESSON OBJECTIVE

Students will describe the forces that cause stars to age and die.

Classroom Activities

On Your Mark!

Prepare for class by checking out 10–20 reference books containing information about the different kinds of stars in the universe (e.g., **neutron stars** also called **pulsars**, **black holes**, etc.).

Review the forces of gravity and nuclear fusion that cause a star to be born, age, collapse, explode, and be transformed into another kind of star. Yellows stars like our sun become red giants. Red giants become white dwarfs, neutron stars, or black holes. Use the information in the Teacher's Classwork Agenda and Content Notes as well as the student Fact Sheet to discuss the different fates of stars of varying masses.

Get Set!

Explain that the velocity an object must have to escape the gravitational attraction of a very massive object depends upon the mass and radius of the very massive object. To escape the gravitational pull of earth, for example, a rocket must achieve a speed of more than 40,323 kilometers per hour. When a large star collapses its mass might be so intense that an object trying to pull free of it would have to reach a velocity of 300,000 kilometers per second. This is the speed of light. According to **Albert Einstein**'s (b. 1879; d. 1955) **special theory of relativity** nothing in the universe can go faster than light! Scientists believe that any star larger than 2.5 times the mass of our sun will eventually collapse into an object so massive that not even light could escape from its surface. If this is so, then the object would not emit light. It would be black! A black hole. Point out that scientists do not yet completely understand how black holes work or what happens to the matter that gets pulled into them. However, several stars in our galaxy have been identified as the companions of massive black holes. Using radio telescopes to monitor the energy being radiated by these stars, scientists can watch the material from them being drawn off at incredible rates of speed. The only object capable of "stripping" these stars of their matter would be a black hole.

Go!

Give students ample time to gather information about the neutron stars (e.g., pulsars) and black holes described in the reference materials you provide.

Materials

library references

EA15 JOURNAL SHEET #3

STARS AND GALAXIES OF THE COSMOS

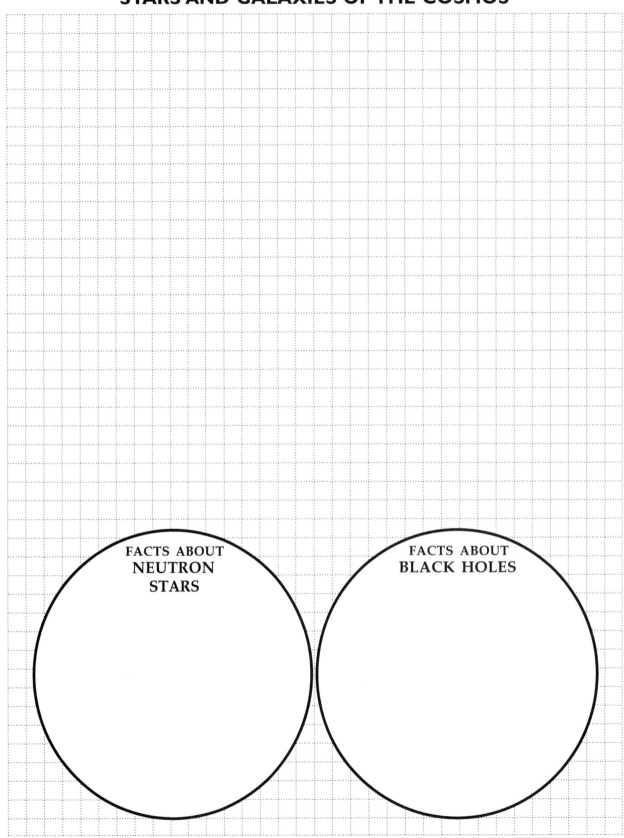

FACTS ABOUT
**NEUTRON
STARS**

FACTS ABOUT
BLACK HOLES

STARS AND GALAXIES OF THE COSMOS

Work Date: ____/____/____

LESSON OBJECTIVE

Students will examine evidence that the universe is expanding.

Classroom Activities

On Your Mark!

Use a felt tip marker to draw dots on a large balloon. Blow up the balloon at the start of class and show students how all of the dots move away from one another as the balloon expands.

Use the information in the Teacher's Classwork Agenda and Content Notes and student Fact Sheet to introduce the work of American physicist **Edwin Powell Hubble** (b. 1889; d. 1953). According to Hubble's observations—and the observations of thousands of astronomers since—all of the galaxies appear to be speeding away from one another like dots on an inflating balloon. For this to be happening, the galaxies must have been given an initial "push"! This is the idea behind the **Big Bang Theory**.

Get Set!

Ask: "What force might prevent the universe from expanding forever?" Answer: gravity. Ask: "What determines the amount of gravity in a system?" Answer: the amount of mass in the system. Explain that scientists have calculated the approximate amount of mass in the universe based on the estimated numbers of stars, galaxies, and nebulae. And they have come up with a very strange answer. There doesn't appear to be enough matter in the universe to stop the expansion! Scientists call this type of universe an **open universe**. Ask: "Is it possible that the 'extra matter' needed to help gravity pull the universe back together exists but that we just can't see it?" Answer: Yes. Scientists are still looking for this mysterious **dark matter**. A universe that expands, then collapses, is called a **closed universe**.

Go!

Give students ample time to complete the activity described in Figure C on Journal Sheet #4. They will discover that the perimeter of the triangle more than doubles. This is because the surface area of a sphere increases as the square of the radius. Point out that the "surface area" of the universe more than doubles as its diameter doubles.

Materials

felt tip markers, large balloons, metric rulers, string

EA15 Journal Sheet #4

STARS AND GALAXIES OF THE COSMOS

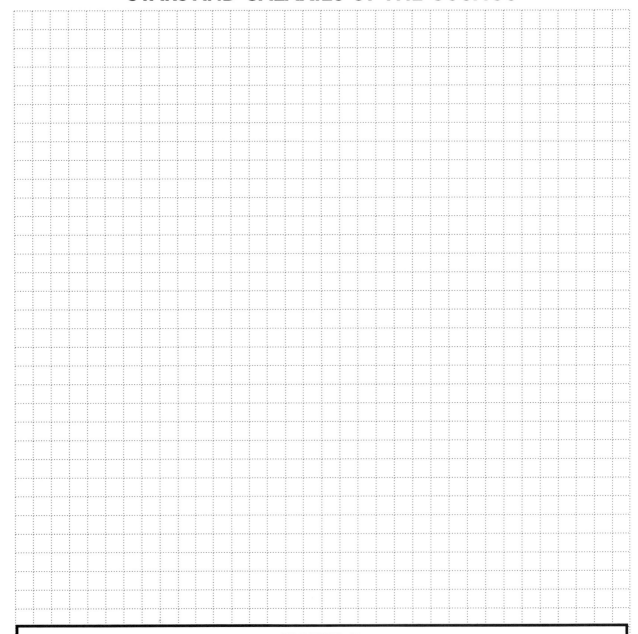

FIGURE C

<u>Directions</u>: (1) Use a felt tip marker to draw 3 dots on a balloon. (2) Connect the dots to form a triangle. (3) Measure the perimeter of the triangle using a length of string and a ruler. Place the string over each side of the triangle; then measure that side length against a ruler. Add up the three sides to get the total perimeter. (4) Blow up the balloon to a diameter of 10 centimeters. (5) Measure and record the perimeter of the triangle. (6) Blow up the balloon to a diameter of 20 centimeters so that the diameter of the balloon is doubled. (7) Measure and record the perimeter of the triangle again in the same way you did in step #3. (8) By dividing the perimeter of the last triangle by the perimeter of the previous triangle you can see how many times the triangle has grown. Did it more than double? Explain your conclusion. What does this tell you about how the distance between galaxies is increasing as the universe doubles and triples in diameter?

EA15 REVIEW QUIZ

Directions: Keep your eyes on your own work.
Read all directions and questions carefully.
THINK BEFORE YOU ANSWER!
Watch your spelling, be neat, and do the best you can.

CLASSWORK	(~40): _____
HOMEWORK	(~20): _____
CURRENT EVENT	(~10): _____
TEST	(~30): _____
TOTAL	(~100): _____

(A ≥ 90, B ≥ 80, C ≥ 70, D ≥ 60, F < 60)

LETTER GRADE: _____

TEACHER'S COMMENTS: _____

_____

STARS AND GALAXIES OF THE COSMOS

TRUE–FALSE FILL-IN: If the statement is true, write the word TRUE. If the statement is false, change the underlined word to make the statement true. *10 points*

_____ 1. <u>Cosmetology</u> is study of the universe.

_____ 2. In 1930, American physicist Edwin Hubble reported that the universe is <u>contracting</u>.

_____ 3. In 1948, American physicist George Gamow proposed the <u>Big Crunch Theory</u> to explain how the universe was born.

_____ 4. At one time all the matter and energy in the universe may have been squeezed into a very small amount of space called a <u>singularity</u>.

_____ 5. The universe is made mostly of <u>oxygen</u> and helium.

_____ 6. Elements lighter than iron are created during nuclear <u>fission</u> inside stars.

_____ 7. Whether or not the universe eventually collapses depends on how much <u>space</u> there is in the cosmos.

_____ 8. There are <u>more</u> galaxies in the universe than there are stars in the Milky Way.

_____ 9. The balanced forces holding a star together are nuclear fusion and <u>gravity</u>.

_____ 10. Best estimates suggest that the universe is about 15 <u>thousand</u> years old.

EA15 Review Quiz *(cont'd)*

MATCHING: Choose the letter of the object that best fits the description. *20 points*

_____ 11. hydrogen fusion balanced by gravity	(A)	red giant
_____ 12. spins extremely fast while pulsing out radio energy	(B)	nova
_____ 13. billions of stars in a group	(C)	neutron star
_____ 14. gathers radio waves for study	(D)	black hole
_____ 15. interstellar dust and gas	(E)	optical telescope
_____ 16. many galaxies	(F)	supercluster
_____ 17. densest known object	(G)	new star
_____ 18. gathers visible light for study	(H)	radio telescope
_____ 19. star that fuses helium to form carbon	(I)	galaxy
_____ 20. exploding star	(J)	nebula

APPENDIX

NAME: _____ **PERIOD:** _____ **DATE:** __ / __ / __

Keep this Grade Roster in the Science Section of your notebook

Date	Journal Points	Homework Points	Current Events Points	Quiz Points	Total Points	Letter Grade	Initials

How to Calculate Your Grade Point Average

Your Report Card grades in this class will be awarded to you according to your grade point average or GPA. You can calculate your GPA whenever you like to find out exactly how you are doing in this class.

First, award each of your weekly grades the following credits: each A is worth 4 credits; each B is worth 3 credits; each C is worth 2 credits; each D is worth 1 credit; and each F is worth 0.

Add your total credits earned. Then, divide by the number of packets listed on your Grade Roster and round the decimal result to the nearest tenths place. Your overall Letter Grade is assigned according to the following GPA values:

A+ ≥ 4.0	A ≥ 3.7	A– ≥ 3.4
B+ ≥ 3.1	B ≥ 2.8	B– ≥ 2.5
C+ ≥ 2.2	C ≥ 1.9	C– ≥ 1.6
D+ ≥ 1.3	D ≥ 1.0	D– ≥ 0.7
	F < 0.7	

FOR EXAMPLE:

John has completed five weeks of school and entered his grades from five packets on his Grade Roster. His grades are as follows: first week, A; second week, B; third week, C; fourth week, C; and fifth week, D.

John awards himself the correct amount of credit for each of his grades.

A	earns	4 credits
B	earns	3 credits
C	earns	2 credits
C	earns	2 credits
D	earns	1 credit
Total	earned is	12 credits

John divides his total credits earned by 5 (the number of packets on his Grade Roster).

12 divided by 5 equals 2.4

John's grade point average, or GPA, is 2.4. Referring to the grades shown above, John knows that he has a C+ in Science thus far, because 2.4 is greater than 2.2 (C+) but less than 2.5 (B–).

Name: _____ **Period:** _____ **Date:** _____ / _____ / _____

Extra Journal Sheet

Fact Sheet Title: _____ **Lesson #** _____

USING CURRENT EVENTS TO INTEGRATE SCIENCE INSTRUCTION ACROSS CONTENT AREAS

Science does not take place in a vacuum. Scientists, like other professionals, are influenced by the economic and political realities of their time. In addition, the ideological and technological advances made by science can influence the economic and political structure of society—for better or worse. It is therefore essential that students have an awareness of the day-to-day science being done at laboratories around the world, important work being reported by an international news media.

Most State Departments of Education, make **CURRENT EVENTS** a regular part of their state science frameworks. Science instructors can use newspaper, magazine, and television reports to keep their students informed about the advances and controversies stemming from research in the many scientific disciplines. Teachers can also use current events to integrate science instruction across the curriculum.

Set aside a class period to show students how to prepare a **science** or **technology** current event. They can do this on a single sheet of standard looseleaf. You may require pupils to read all or part of a science/technology article depending on their reading level. Have them practice summarizing the lead and one or more paragraphs of the article *in their own words*. Advise them to keep a **thesaurus** on hand or to use the dictionary/thesaurus stored in their personal computer at home. Tell students to find *synonyms* they can use to replace most of the vocabulary words used by the article's author. This activity will help them to expand their vocabulary and improve grammar skills. Show students how to properly trim and paste the article's title and first few paragraphs on the front of a standard piece of looseleaf. They should write their summary on the opposite side of the page so that the article is visible to their classmates when they present their findings orally to the class. Allow students to make a report that summarizes a newsworthy item they may have heard on television. The latter report should be accompanied by the signature of a parent/guardian to insure the accuracy of the information being presented.

Students' skills at public speaking are sure to improve if they are given an opportunity to share their current event. Current events can be shared after the end-of-the-unit REVIEW QUIZ or whenever the clock permits at the end of a lesson that has been completed in a timely fashion. You can select students at random to make their presentations by drawing lots or ask for volunteers who might be especially excited about their article. Take time to discuss the ramifications of the article and avoid the temptation to express your personal views or bias. Remain objective and give students the opportunity to express their views and opinions. Encourage them to base their views on fact, not superstition or prejudice. Should the presentation turn into a debate, set aside a few minutes later in the week, giving students time to prepare what they would like to say. Model courtesy and respect for all points of view and emphasize the proper use of the English language in all modes of presentation, both written and oral.

BIO-DATA
CARDS

BIO-DATA CARD

JOHN COUCH ADAMS
(born 1819; died 1892)

nationality
English

contribution to science
predicted the existence of Neptune based on variations in the orbit of Uranus

BIO-DATA CARD

BUZZ ALDRIN
(born 1930)

nationality
American

contribution to science
second man to walk on the moon in July, 1969

BIO-DATA CARD

LUIS ALVAREZ
(born 1911; died 1988)

nationality
American

contribution to science
first to suggest that a meteor collision with earth caused the extinction of the dinosaurs

BIO-DATA CARD

ARCHIMEDES
(born 287 B.C.; died 212 B.C.)

nationality
Italian

contribution to science
discovered the principle of buoyancy laying the foundation for the development of the submersible

BIO-DATA CARD

ARISTOTLE
(born 384 B.C.; died 322 B.C.)

nationality
Greek

contribution to science
observations served as the basis of scientific theory until the middle of the 15th century

BIO-DATA CARD

NEIL ARMSTRONG
(born 1930)

nationality
American

contribution to science
first man to walk on the moon in July, 1969

BIO-DATA CARD

ROBERT D. BALLARD
(born 1942)

nationality
American

contribution to science
discovered the sunken wreck of the luxury liner Titanic in 1985

BIO-DATA CARD

FRANCIS BEAUFORT
(born 1774; died 1857)

nationality
French

contribution to science
established the first wind velocity scale to measure the effects of the wind

BIO-DATA CARD

VILHELM F.K. BJERKNES
(born 1862; died 1951)

nationality
Norwegian

contribution to science
considered the "father of modern meteorology" for describing the movement of air masses

BIO-DATA CARD

JOHANN ELERT BODE
(born 1747; died 1826)

nationality
German

contribution to science
derived the first useful law to help predict the location of planets, although the law later proved to be wrong

BIO-DATA CARD

TYCHO BRAHE
(born 1546; died 1601)

nationality
Danish

contribution to science
made voluminous records about planetary motion that led to the heliocentric theory of the universe

BIO-DATA CARD

WERNER von BRAUN
(born 1912; died 1977)

nationality
German

contribution to science
assisted American scientists in the development of multistage rockets capable of going into space

BIO-DATA CARD

ALEXANDRE BRONGNIART
(born 1770; died 1837)

nationality
French

contribution to science
with Cuvier developed the first system for classifying fossils according to their anatomical similarities

BIO-DATA CARD

ROBERT BUNSEN
(born 1811; died 1899)

nationality
German

contribution to science
with Kirchoff invented the first spectroscope

BIO-DATA CARD

ANDERS CELSIUS
(born 1701; died 1744)

nationality
Swedish

contribution to science
introduced the centigrade temperature scale in 1742

BIO-DATA CARD

NICOLAUS COPERNICUS
(born 1473; died 1543)

nationality
Polish

contribution to science
first to propose that the sun was the center of the known universe and not the earth

INSTRUCTIONS TO TEACHERS
Xerox and cut out the Bio-Data Cards below and keep them in a handy file. Instruct students to choose one card and neatly glue it to the front of a 5″ × 8″ index card. They can use the school or public library to find out more about the scientist they have chosen. On the back of the index card they can draw a cartoon, write a poem or short paragraph that illustrates an important event in the life of this famous personality.

BIO-DATA CARD

GASPARD GUSTAVE de CORIOLIS
(born 1792; died 1843)

nationality
French

contribution to science
explained the rotational movement of hemispheric air masses today called the Coriolis effect

BIO-DATA CARD

NICHOLAS de CUSA
(born 1401; died 1464)

nationality
German

contribution to science
invented the first hygroscope to measure the humidity of the air

BIO-DATA CARD

GEORGES CUVIER
(born 1769; died 1832)

nationality
French

contribution to science
with Brongniart developed the first system for classifying fossils according to their anatomical similarities

BIO-DATA CARD

CHARLES DARWIN
(born 1809; died 1882)

nationality
English

contribution to science
proposed the theory of evolution by means of natural selection to explain the diversity of life on earth

BIO-DATA CARD

WILLIAM MORRIS DAVIS
(born 1850; died 1934)

nationality
American

contribution to science
proposed the erosion cycle to explain the diversity of rocks and landforms

BIO-DATA CARD

RENÉ DESCARTES
(born 1596; died 1650)

nationality
French

contribution to science
proposed a new "mechanical philosophy" to explain natural phenomena

BIO-DATA CARD

CORNELIS DREBBEL
(born 1572; died 1633)

nationality
Dutch

contribution to science
invented the first successful oar-propelled submarine

BIO-DATA CARD

ALBERT EINSTEIN
(born 1879; died 1955)

nationality
German-American

contribution to science
discovered a basic law of physics that explained the relationship between energy, mass, and the speed of light

BIO-DATA CARD

ERATOSTHENES
(born 276 B.C.; died 194 B.C.)

nationality
Greek

contribution to science
made the first accurate estimate of earth's circumference by comparing the angles created by shadows

BIO-DATA CARD

EUCLID
(born 330 B.C.; died 260 B.C.)

nationality
Greek

contribution to science
invented geometry to help explain the patterns that occurred in nature

BIO-DATA CARD

GABRIEL DANIEL FAHRENHEIT
(born 1686; died 1736)

nationality
Dutch

contribution to science
invented the Fahrenheit thermometer in 1718

BIO-DATA CARD

JEAN BERNARD LÉON FOUCAULT
(born 1819; died 1868)

nationality
French

contribution to science
constructed a pendulum that detected the earth's rotation on its axis

BIO-DATA CARD

JOSEPH von FRAUNHOFER
(born 1787; died 1826)

nationality
German

contribution to science
suggested that the spectrum of light produced by the sun might give clues to the sun's chemical composition

BIO-DATA CARD

YURI GAGARIN
(born 1938; died 1968)

nationality
Russian

contribution to science
first man to orbit the earth

BIO-DATA CARD

GALILEO GALILEI
(born 1564; died 1642)

nationality
Italian

contribution to science
popularized the heliocentric theory of the universe and discovered the rings of Saturn

BIO-DATA CARD

JOHANN GOTTFRIED GALLE
(born 1812; died 1910)

nationality
German

contribution to science
discovered the planet Neptune in 1846

BIO-DATA CARD
GEORGE GAMOW
(born 1904; died 1968)

nationality
American

contribution to science
popularized the Big Bang theory of the universe

BIO-DATA CARD
HANS GEIGER
(born 1882; died 1945)

nationality
German

contribution to science
invented the Geiger counter to measure the rate of radioactive disintegration in rocks

BIO-DATA CARD
ROBERT HUTCHINGS GODDARD
(born 1882; died 1945)

nationality
American

contribution to science
launched the first liquid fueled rocket

BIO-DATA CARD
BENO GUTENBERG
(born 1889; died 1960)

nationality
German

contribution to science
determined the outer boundary of earth's core to be 2,900 kilometers beneath the surface of the crust

BIO-DATA CARD
GEORGE HADLEY
(born 1685; died 1768)

nationality
English

contribution to science
suggested that earth's rotation governed the motion and direction of the winds

BIO-DATA CARD
EDMOND HALLEY
(born 1656; died 1742)

nationality
English

contribution to science
discovered the comet that bears his name and suggested an explanation for the trade winds that blow near the equator

BIO-DATA CARD
JOHN HARRISON
(born 1693; died 1776)

nationality
English

contribution to science
invented the first marine chronometer in 1735

BIO-DATA CARD
RENÉ-JUST HAÜY
(born 1791; died 1867)

nationality
English

contribution to science
developed the first classification system used to classify crystals

---

BIO-DATA CARD

JOSEPH HENRY
(born 1797; died 1878)

nationality
American

contribution to science
meteorological studies at the
Smithsonian Institute led to the
founding of the US Weather Bureau

BIO-DATA CARD

SIR WILLIAM HERSCHEL
(born 1738; died 1822)

nationality
English

contribution to science
discovered the planet Uranus in 1781
using a 7-inch reflecting telescope of his
own construction

BIO-DATA CARD

EJNAR HERTZSPRUNG
(born 1873; died 1957)

nationality
Danish

contribution to science
with Russell correlated the luminosity
of stars with the
star's surface temperature

BIO-DATA CARD

HIPPARCHUS
(born 190 B.C.; died 120 B.C.)

nationality
Greek

contribution to science
invented the astrolabe to measure
declination above the horizon

BIO-DATA CARD

ARTHUR HOLMES
(born 1890; died 1965)

nationality
English

contribution to science
pioneered radioactive dating methods

BIO-DATA CARD

ROBERT HOOKE
(born 1635; died 1703)

nationality
English

contribution to science
promoted a "mechanical philosophy"
to explain the laws governing
the actions of nature

BIO-DATA CARD

EDWIN POWELL HUBBLE
(born 1889; died 1953)

nationality
American

contribution to science
proved that galaxies were groups of stars
outside the Milky Way and showed that
the universe is expanding

BIO-DATA CARD

JAMES HUTTON
(born 1726; died 1797)

nationality
Scottish

contribution to science
known as the "father of geology" for his
explanation of the principle of uniformity

BIO-DATA CARD
IMMANUEL KANT
(born 1724; died 1804)

nationality
German

contribution to science
popularized the idea that the earth was formed from a cloud of dust and gas

BIO-DATA CARD
WILLIAM THOMSON KELVIN
(born 1824; died 1907)

nationality
Irish

contribution to science
invented a temperature scale used for measuring extremely cold objects

BIO-DATA CARD
JOHANNES KEPLER
(born 1571; died 1630)

nationality
German

contribution to science
proved that the planets revolved around the sun in elliptical orbits

BIO-DATA CARD
GUSTAV ROBERT KIRCHOFF
(born 1824; died 1887)

nationality
German

contribution to science
with Bunsen invented the first practical spectroscope

BIO-DATA CARD
WLADMIR PETER KÖPPEN
(born 1846; died 1940)

nationality
German

contribution to science
created the first system of classification to describe the earth's climatic zones

BIO-DATA CARD
PAUL LANGEVIN
(born 1872; died 1946)

nationality
French

contribution to science
first to use sonar—sound navigation and ranging—to determine the depths of the ocean

BIO-DATA CARD
HENRIETTA SWAN LEAVITT
(born 1868; died 1921)

nationality
American

contribution to science
discovered the period-luminosity law of variable stars allowing a more accurate determination of stellar distances

BIO-DATA CARD
INGE LEHMANN
(published in 1936)

nationality
Danish

contribution to science
discovered the inner core of the earth

BIO-DATA CARD
GEORGES EDOUARD LEMAÎTRE
(born 1894; died 1996)

nationality
Belgian

contribution to science
proposed the Big Bang theory of the universe in 1927

BIO-DATA CARD
HANS LIPPERSHEY
(born 1570; died 1619)

nationality
Danish

contribution to science
invented the refracting telescope in 1608

BIO-DATA CARD
PERCIVAL LOWELL
(born 1855; died 1916)

nationality
American

contribution to science
predicted the existence of the planet Pluto

BIO-DATA CARD
CHARLES LYELL
(born 1797; died 1875)

nationality
Scottish

contribution to science
made the first serious estimate of earth's age arguing that it was millions of years older than most people thought

BIO-DATA CARD
GERHARD KREMER MERCATOR
(born 1512; died 1594)

nationality
Flemish

contribution to science
published the first map showing accurately plotted points taken from a curved surface

BIO-DATA CARD
MILUTIN MILANKOVICH
(born 1879; died 1958)

nationality
Yugoslavian

contribution to science
suggested a widely accepted hypothesis to explain the periodic recurrence of Ice Ages

BIO-DATA CARD
JOHN MILNE
(born 1850; died 1913)

nationality
English

contribution to science
invented the first pendulum seismometer

BIO-DATA CARD
FRIEDRICH MOH
(born 1773; died 1839)

nationality
German

contribution to science
created the first hardness scale to assist in the identification and classification of rocks and minerals

BIO-DATA CARD

Joseph Michel & Jacques Etienne MONTGOLFIER

(born 1740; died 1810)   (born 1745; died 1799)

nationality
French

contribution to science
made the first successful flight
in a hot air balloon

BIO-DATA CARD

SAMUEL F. B. MORSE

(born 1791; died 1872)

nationality
American

contribution to science
invented the telegraph in 1838 that
made possible the first large scale
weather maps

BIO-DATA CARD

JOHN MURRAY

(born 1841; died 1914)

nationality
English

contribution to science
authored 50 volumes of oceanographic
maps and descriptive text

BIO-DATA CARD

SIR ISAAC NEWTON

(born 1642; died 1727)

nationality
English

contribution to science
derived three laws of motion and a
universal law of gravity to explain the
forces that moved the planets

BIO-DATA CARD

RICHARD DIXON OLDHAM

(born 1858; died 1936)

nationality
English

contribution to science
identified the three basic types of
seismic waves: primary, secondary, and
surface waves

BIO-DATA CARD

ABRAHAM ORTELIUS

(born 1527; died 1598)

nationality
Danish

contribution to science
published the first modern
world atlas

BIO-DATA CARD

Auguste Antoine & Jacques PICCARD

(born 1884; died 1962) (born 1922)

nationality
Swiss

contribution to science
father and son who designed and
submerged the first deep sea bathyscaphe

BIO-DATA CARD

WILLIAM HENRY PICKERING

(born 1858; died 1938)

nationality
American

contribution to science
with Tombaugh found
the planet Pluto in 1930

BIO-DATA CARD

PLATO
(born 427 B.C.; died 347 B.C.)

nationality
Greek

contribution to science
philosopher whose logical arguments
allowed scientists to develop the
scientific method

BIO-DATA CARD

JOSEPH PRIESTLEY
(born 1733; died 1804)

nationality
English

contribution to science
discovered nitrogen—the gas that
comprises 78% of the earth's
atmosphere

BIO-DATA CARD

CLAUDIUS PTOLEMAEUS
(born 100 A.D.; died 170 A.D.)

nationality
Egyptian

contribution to science
made accurate maps of Europe, North
Africa, and the Middle East and developed
the geocentric theory of the universe

BIO-DATA CARD

WILLIAM REDFIELD
(born 1789; died 1857)

nationality
American

contribution to science
identified the origin of tropical storms

BIO-DATA CARD

SALLY KIRSTEN RIDE
(born 1951)

nationality
American

contribution to science
first woman to ride
the space shuttle into orbit

BIO-DATA CARD

HENRY NORRIS RUSSELL
(born 1877; died 1957)

nationality
American

contribution to science
with Hertzsprung showed the
correlation between a star's luminosity
and surface temperature

BIO-DATA CARD

KARL WILHELM SCHEELE
(born 1742; died 1786)

nationality
Swedish

contribution to science
shared recognition with Priestley for
the discovery of oxygen—the gas that
comprises 21% of earth's atmosphere

BIO-DATA CARD

ADAM SEDGWICK
(born 1785; died 1873)

nationality
English

contribution to science
named the oldest strata of fossils, the
Cambrian layer, dated at 590 to 500
million years old

BIO-DATA CARD

ALAN BARTLETT SHEPARD
(born 1923; died 1998)

nationality
American

contribution to science
first American to fly in space

BIO-DATA CARD

WILLIAM SMITH
(born 1769; died 1839)

nationality
English

contribution to science
introduced the idea of the index fossil to aid in the dating of fossil-containing rock strata

BIO-DATA CARD

NIKOLAUS STENO
(born 1638; died 1686)

nationality
Danish

contribution to science
used fossil evidence to show that the earth's crust had been deposited in layers

BIO-DATA CARD

VALENTINA V. TERESHKOVA
(born 1937)

nationality
Russian

contribution to science
first woman to orbit the earth

BIO-DATA CARD

JOHANN DANIEL TITIUS
(born 1747; died 1796)

nationality
German

contribution to science
helped derive the first useful law to predict the location of planets, although the law later proved to be wrong

BIO-DATA CARD

CLYDE WILLIAM TOMBAUGH
(born 1906; died 1997)

nationality
American

contribution to science
with Pickering discovered the planet Pluto in 1930

BIO-DATA CARD

EVANGELISTA TORRICELLI
(born 1608; died 1647)

nationality
Italian

contribution to science
invented the barometer to measure air pressure and proved the existence of a vacuum

BIO-DATA CARD

KONSTANTIN E. TSIOLKOVSKY
(born 1857; died 1935)

nationality
Russian

contribution to science
calculated the escape velocity of earth and proposed the first feasible theory of spaceflight

BIO-DATA CARD
JAMES ALFRED VAN ALLEN
(born 1914)

nationality
American

contribution to science
discovered the radiation belts
surrounding the earth

BIO-DATA CARD
JULES VERNE
(born 1828; died 1905)

nationality
French

contribution to science
popularized the notion of exploration
and space travel by authoring scores
of science fiction stories and novels

BIO-DATA CARD
ALFRED LOTHAR WEGENER
(born 1880; died 1930)

nationality
German

contribution to science
proposed the theory of continental drift
later called the theory of plate tectonics

BIO-DATA CARD
ABRAHAM GOTTLOB WERNER
(born 1749; died 1817)

nationality
German

contribution to science
established the first physical explanation
for the process of geological stratification

BIO-DATA CARD
FRED LAWRENCE WHIPPLE
(born 1906)

nationality
American

contribution to science
proposed the dust-cloud hypothesis of
solar system formation a revision of
Kant's nebular hypothesis

BIO-DATA CARD
WILLIAM WOLLASTON
(born 1766; died 1828)

nationality
English

contribution to science
recorded the first solar spectrum in 1802

BIO-DATA CARD
Orville & Wilbur WRIGHT
(born 1871; died 1948)   (born 1867; died 1912)

nationality
American

contribution to science
made the first successful flight of a
gasoline powered airplane in 1903